Probability and Randomness
Quantum versus Classical

Probability and Randomness
Quantum versus Classical

Andrei Khrennikov
Linnaeus University, Sweden

Imperial College Press

Published by

Imperial College Press
57 Shelton Street
Covent Garden
London WC2H 9HE

Distributed by

World Scientific Publishing Co. Pte. Ltd.
5 Toh Tuck Link, Singapore 596224
USA office: 27 Warren Street, Suite 401-402, Hackensack, NJ 07601
UK office: 57 Shelton Street, Covent Garden, London WC2H 9HE

Library of Congress Cataloging-in-Publication Data
Names: Khrennikov, A. Yu. (Andrei Yurievich), 1958–   author.
Title: Probability and randomness: quantum versus classical / by Andrei Khrennikov
   (Linnaeus University, Sweden).
Description: Covent Garden, London : Imperial College Press, [2016] | Singapore ;
   Hackensack, NJ : Distributed by World Scientific Publishing Co. Pte. Ltd. | 2016 |
   Includes bibliographical references and index.
Identifiers: LCCN 2015051306| ISBN 9781783267965 (hardcover ; alk. paper) |
   ISBN 1783267968 (hardcover ; alk. paper)
Subjects: LCSH: Probabilities. | Quantum theory. | Mathematical physics.
Classification: LCC QC20.7.P7 K47 2016 | DDC 530.13--dc23
LC record available at http://lccn.loc.gov/2015051306

British Library Cataloguing-in-Publication Data
A catalogue record for this book is available from the British Library.

Printed in Singapore

To my son Anton

Preface

The education system for physics students worldwide suffers from the absence of a deep course in probability and randomness. This is a real problem for students interested in quantum information theory, quantum optics, and quantum foundations. Here a primitive treatment of probability and randomness may lead to deep misunderstanding of the theory and wrong interpretations of experimental results. Since my visits (in 2013 and 2014 by kind invitations of C. Brukner and A. Zeilinger) to the Institute for Quantum Optics and Quantum Information (IQOQI) of Austrian Academy of Sciences, a number of students (experimentalists!) have been asking me about foundational problems of probability and randomness, especially inter-relation between classical and quantum structures. I gave two lectures on these problems [165]. Surprisingly, experiment-oriented students demonstrated very high interest in mathematical peculiarities. This (as well as frequent reminder of Prof. Zeilinger) motivated me to write a text based on these lectures which were originally presented in the traditional black-board form. The main aim of this book is to provide a short foundational introduction to classical and quantum probability and randomness.

Chapter 1 starts with the presentation of the Kolmogorov (1933) *measure-theoretic axiomatics*. The von Mises frequency probability theory which preceded the Kolmogorov theory is also briefly presented.[1] In this chapter we discuss interpretations of probability notable for their diversity which is similar to the diversity of interpretations of a quantum state.

---

[1]Now this theory is practically forgotten. However, it played an important role in search for an adequate axiomatics of probability theory and randomness, especially von Mises' *principle of randomness*. We proceed with the Kolmogorov theory, see my monographs [133], [156] for von Mises probability versus quantum probability.

Already in Chapter 1 we derive a version of the famous *Bell inequality* [31] (in the Wigner form) as expressing two basic properties of a measure: additivity and non-negativity. The derivation is based on the assumption of the existence of *a single probability measure* serving to represent all probability distributions involved in this inequality. We remark that Kolmogorov endowed his model of probability with a "protocol" of its application: each complex of experimental conditions (i.e., each context) is described by its own probability space. Thus in any multi-contextual experiment, such as experiments on Bell's inequality, we are dealing, in general, with a family of probabilities corresponding to different contexts. Kolmogorov studied the problem of the existence of the common probability space for stochastic processes and found the corresponding necessary and sufficient conditions.

Chapter 3 may be difficult for physicists. Here we present the standard construction of the *Lebesgue extension* of a countably additive measure which is originally defined on a simple system of sets. An example of *non-measurable set* is of foundational interest. It contradicts (physical) intuition that probability can be assigned to any event.[2] Moreover, its existence is based the *axiom of choice* (E. Zermelo, 1904). The formulation of this axiom taken by itself sounds still acceptable. However, it has some equivalent formulations, e.g., one known as "the well-ordering theorem", which are really counter-intuitive. Some mathematicians are suspicious of this axiom. Our aim was to show that the foundations of classical (measure-theoretic) probability are more ambiguous than the foundations of quantum (complex Hilbert space) probability. The last section of Chapter 3 presents "exotic generalization of concept of probability" such as *negative probability* (cf. Dirac, Feynman, Aspect) and *p*-adic probability with possible applications in quantum foundations.

Chapter 4 contains the basics of the quantum formalism. This chapter plays the introductory role for a newcomer to quantum theory, but it can also be interesting for physicists. Here we proceed by using general theory of *quantum instruments*.

Chapter 5 gets to the core of classical versus quantum probability interplay. It starts with Feynman's analysis of the probability structure of the *two-slit experiment* [88]. His conclusion is that classical probability is not applicable to results of the multi-contextual structure of this experiment.

---

[2]This non-measurability argument was explored by I. Pitowsky in his analysis of violation of Bell's inequality [218]. However, nowadays it is completely forgotten. Nobody would say: "Bell's inequality is violated because some sets of hidden variables are not measurable."

The rest of the chapter is devoted to the mathematical formalization of this contextual probability viewpoint.[3]

In my previous publications, e.g., [156], I treated contextual probability as *non-Kolmogorovian probability*, by following Accardi (see [2] - [4] - unfortunately, he was not able to publish his book in English).[4] However, now I understand that this terminology has to be used with caution. Formally, probabilistic data from two-slit experiment and Bell's inequality experiment can be embedded in the classical probability space. However, this embedding is not straightforward: probabilities have to be treated as conditional, see Chapter 8 for construction of this embedding for Bell's experiment.

For me, Bell's argument sounds as follows: *we cannot represent probabilities collected for different pairs of orientations of polarization beam splitters as unconditional classical probabilities.* However, I am not sure that Bell would accept this interpretation. He was concentrated on the nonlocality dimension - by trying to justify Bohmian mechanics as the genuine quantum model.[5]

Chapter 6 is the most difficult for reading. It is about interpretations of quantum mechanics. The main problem is their diversity. My attempt to classify them may be found boring. One can just scan this chapter: the classical interpretations of von Neumann and Einstein-Ballentine and modern ones such as the information interpretation (Zeilinger-Brukner), statistical Copenhagen interpretation (Plotnitsky), QBism (Fuchs, Schack, Mermin), and the Växjö interpretation (Khrennikov). Zeilinger, Brukner and Plotnitsky can be considered as neo-Copenhagenists. QBists are also often treated in the same way. However, this is the wrong viewpoint on QBism. The Växjö interpretation can be considered as merging the Einstein-Ballentine ensemble interpretation and Bohr's contextual viewpoint on quantum

---

[3]We remark that R. Feynman appealed to the two-slit experiment in all his discussions on quantum foundations. Similarly to N. Bohr, he considered this experiment as the heart of quantum mechanics (we remark that the same point was permanently expressed in publications and talks of L. Accardi). I share this viewpoint of Bohr-Feynman-Accardi. On the other hand, nowadays one can often hear that entanglement and violation of Bell's inequality (rather than interference demonstrated in the two-slit experiment) are the key elements of quantum theory. I do not think so as from the contextual viewpoint the two-slit and Bell experiments are of the same nature. Mathematically both are expressed as violations of theorems of Kolmogorovian probability theory: the *formula of total probability* (the two-slit experiment) and the Bell inequality.

[4]Feynman [88] did not hear about Kolmogorov's axiomatics of probability theory; he wrote about violation of laws of Laplacian probability theory.

[5]Chapter 8 may do harm for a young physicist's attitude towards the nonlocality problem. If the idea of quantum nonlocality is dear to the reader, probably just skip this chapter.

observables. Surprisingly, this interpretation has a lot in common with QBism (see Chapter 6).

Chapter 7 is devoted to quantum randomness. This is mainly a philosophic discussion about inter-relation of von Neumann's irreducible randomness and classical approaches to randomness. Finally, we discuss possible applications of quantum probability outside of physics, cognition, psychology, biology, economics, Chapter 9. This chapter is of introductory character and its aim is just to inform the reader about such applications.

I hope that the book will serve as a textbook on classical and quantum probability and randomness (Chapters 1, 2, 4, 5, 7) and interpretations of quantum mechanics (Chapter 6). Chapter 8 presents the author's viewpoint on Bell's inequality and Chapter 9 informs readers about new areas of application of the quantum formalism: biology, cognition, economics.

This book combines short mathematical introductions to probability and randomness with rather long discussions on their interpretations. Readers who are not so much interested in the latter can simply skip interpretational parts.

I would like to thank I. Basieva, C. Fuchs, A. Plotnitsky, and Zeilinger for numerous critical discussions on interpretational issues. Their views differ crucially from my own (and from each other) and such discussions were fruitful for me. This is a good occasion to thank once again A. Bulinsky and A. Shyryaev who explained to me that Kolmogorov's viewpoint on probability was *contextual*: each complex of experimental conditions determines its own probability space, see section 2 in [177].

Vienna – Växjö, 2013–2015.

Contents

Chapter 1

Foundations of Probability

We start with the remark that, in contrast with, e.g., geometry, axiomatic probability theory was created not so long ago. Soviet mathematician Andrei Nikolaevich Kolmogorov presented the modern axiomatics of probability theory only in 1933 in his book [177]. The book was originally published in German.[1] The English translation [178] was published only in 1952 (and the complete Russian translation [179] of the German version [177] only in 1974).[2] Absence of an English translation soon (when the German language lost its international dimension) led to the following problem. The majority of the probability theory community did not have a possibility to read Kolmogorov. Their picture of the Kolmogorov model was based on its representations in English (and Russian) language textbooks. Unfortunately, in such representations a few basic ideas of Kolmogorov disappeared, since they were considered as philosophical remarks with no direct relevance to mathematics. This is partially correct, but *probability theory is not just mathematics*. It is a physical theory and, as any physical theory, its mathematical formalism has to be endowed with some interpretation. In Kolmogorov's book the interpretation question was discussed in very detail. However, in the majority of mathematical representations of Kolmogorov's approach, the interpretation issue is not enlightened at all. From my viewpoint, one of the main negative consequences of this ignorance was oblivion of *contextuality* of Kolmogorov's theory of probability. Kolmogorov designed his probability theory as a mathematical formalization of random experiments (see also discussion below). For him, each exper-

---

[1]The language and the publisher (Springer) were chosen by a rather pragmatic reason. Springer paid in gold, young Kolmogorov felt the need of money, and gold was valuable even in the Soviet Union.

[2]The first Russian version was published in 1936 [176]. However, it was not identical to the original German version, in some places it was shorten.

1

imental context C generates its own probability space (endowed with its own probability measure). It is practically impossible to find a probability theory textbook mentioning this key interpretational issue of Kolmogorov's theory. We remark that contextuality of classical probability theory plays very important role when classical and quantum probability theories are being compared, see Chapter 5.

Fig. 1.1   Andrei Nikolaevich Kolmogorov

We recall that at the beginning of 20th century probability was considered as a part of mathematical physics and not pure mathematics. In 1900 at the Paris mathematical congress David Hilbert presented the famous list of problems [114] - [116]. The 6th problem is about axiomatization or physical theories:

"Mathematical Treatment of the Axioms of Physics. The investigations on the foundations of geometry suggest the problem: To treat in the same manner, by means of axioms, those physical sciences in which already today mathematics plays an important part; in the first rank are the theory of probabilities and mechanics."

It was not clear what features of nature have to be incorporated in an adequate mathematical model of this physical theory, the theory of probability. (This is one of the main reasons of the so late axiomatization.) Kolmogorov's measure-theoretical representation is based on one of the possible selections of features of statistical natural phenomena. Another great figure in foundations of probability, Richard von Mises, selected other features which led to the frequency theory of probability, the theory of random sequences (collectives) [247] - [249].

Nowadays von Mises theory is practically forgotten and Kolmogorov's theory is booming - in particular, as the result of the recent tremendous development of financial mathematics [233]. In fact, this situation is not merely a consequence of better reflection of statistical physical phenomena in Kolmogorov's approach. My personal opinion is that von Mises probability matches better real experimental situations in physics (starting with the highly natural definition of probability as the limit of frequencies). In particular, in the measure-theoretic approach contextuality is shadowed, since it is not explicitly present in its mathematical structure. It can be found only in the discussion on relation of the theory to experiment [177]. In the frequency model contextuality is explicitly encoded in the notion of collective as generated by an experiment, and the coupling experimental context $\mapsto$ experiment is straightforward.

It is again my personal opinion that Kolmogorov's probability is so popular because it is simpler and its logical structure is clearer than von Mises' theory.[3] People prefer simplicity... In this book we shall try to present both basic approaches to formalization of probability, Kolmogorov [177] (1933) and von Mises [247] (1919). However, since this book is aimed to play the role of an introduction to probability for physicists, more attention will be paid to Kolmogorov theory, because of its essentially wider use in theory and applications. One who wants to know more about the frequency approach to probability and its recent applications, in particular to quantum physics, can read books [133], [156]. Typically experts refer to von Mises theory as suffering from absence of rigorousness. However, this is not correct. The initial definition of random sequence (collective) of von Mises was really presented at the physical level of rigorousness. However, later it was perfectly mathematically formalized by Wald [253] and Church [60], see section 2.1.1. The main difficulty arises from the attempt of von Mises to get "in one" both probability and randomness. And this is really a great problem of modern science which has not yet been solved completely, see

---

[3]However, by going to a deeper level - to set theory axiomatics - we find that simplicity and clearness are illusions, see Chapter 3.

Chapter 2.

For now, we only emphasize that, while the notion of randomness is closely related to the notion of probability, they do not coincide. We can say that the problem of proper mathematical formalization of probability was successfully solved, by Kolmogorov. However, as we shall see in Chapter 2, mathematicians are still unable (in spite of a hundred years of tremendous efforts) to provide a proper formalization of randomness.

It is interesting that the foundations of quantum probability and randomness were set practically at the same time as the foundations of classical probability theory. In 1935 John von Neumann [250] pointed out that classical randomness is fundamentally different from quantum randomness. The first one is *"reducible randomness"*, i.e., it can be reduced to variation of features of systems in an ensemble. It can also be called *ensemble randomness*. The second one is *"irreducible randomness"*, i.e., the aforementioned ensemble reduction is impossible. By von Neumann, quantum randomness is an *individual randomness*, even an individual electron exhibits fundamentally random behavior. Moreover, only quantum randomness is genuine randomness – a consequence of violation of causality at the quantum level, see Chapter 7 for more detail.

1.1   Interpretation Problem in Quantum Mechanics and Classical Probability Theory

We address the interpretation problem of classical probability theory, see [133], [156] for a detailed presentation, in comparison with the interpretation problem of quantum mechanics (QM). The latter is well known and is considered one of foundational problems of QM. The present situation is characterized by the diversity of interpretations. Really, this is unacceptable for a scientific theory. Moreover, these interpretations are not just slight modifications of each other. They differ fundamentally: the Copenhagen interpretation of Bohr[4], Heisenberg, Pauli, the ensemble interpretation of Einstein[5], Margenau, and Ballentine (QM is a version of

---

[4]It is not well-known that originally Niels Bohr was convinced to proceed with the operational interpretation of QM by the Soviet physicist Vladimir Fock, the private communication of Andrei Grib who read the correspondence between Bohr and Fock in which Fock advertised actively the "Copenhagen interpretation". But initially Bohr was not ready to share Fock's viewpoint.

[5]Initially Schrödinger also kept to this interpretation. Nowadays it is practically forgotten that he elaborated his example with (Schrödinger) Cat and Poison just to demonstrate the absurdness of the Copenhagen interpretation. This example was, in fact, just a

classical statistical mechanics with may be very tricky phase space), many worlds interpretation (no comment), ..., the Växjö interpretation, section 6.9 (a generalization of the ensemble interpretation taking into account contextuality of measurements) [143], [147].[6]

Our main message is that *the problem of finding proper interpretations for classical probability and randomness is no less complex.* It is also characterized by a huge diversity. For example, there are measure-theoretic probability, frequency probability, subjective (Bayesian) probability, ... and randomness as unpredictability, randomness as complexity, randomness as typicality,....

It is interesting that nowadays in probability theory the problem of interpretation is practically ignored. One can say that the majority of the probability community proceed under the same slogan as the majority of the quantum community: "shut up and calculate".

Partially this is a consequence of the common treatment of probability theory as a theoretical formalism. Questions of applicability of this formalism are addressed by statisticians. Here the interpretation questions play an important role, may be even more important than in QM. By using different interpretations (e.g., frequency and Bayesian) researchers can come to very different conclusions based on the same statistical data; different interpretations generate different methods of analysis of data, e.g., frequentists operate with confidence intervals estimates and Bayesians with credible intervals estimates. Opposite to QM, it seems that nowadays in probability theory and statistics, nobody expects that a single "really proper interpretation" will be finally created.

The situation with randomness-interpretations is more similar to QM. It is characterized by practically one hundred years of discussions on possible interpretations of randomness. Different interpretations led to different theories of randomness. Numerous attempts to elaborate a "really proper interpretation" of randomness have not yet lead to a commonly acceptable result. Nevertheless, there are still expectations that a new simple and

---

slight modification of Einstein's example with Man and Gun in his letter to Schrödinger. However, finally Schrödinger wrote to Einstein that he found it very difficult if at all possible to explain the interference on the basis of the ensemble interpretation, so he gave up.

[6]We can point to the series of the Växjö conferences on quantum foundations where all possible interpretations were argued for and against during the last 15 years (see, e.g., lnu.se/qtap and lnu.se/qtpa for the last conferences and the book [167] devoted to this series). However, in spite of the generally great value of such foundational debates, the interpretation picture of QM did not become clearer.

unexpected idea will come to life and a rigorous mathematical model based on the commonly acceptable notion of randomness will be created (this was one of the last messages of Andrei Kolmogorov before his death, see the footnote at the very end of Chapter 2).

There is an opinion (not very common) that the interpretation problem of QM simply reflects the interpretation problem of probability and randomness.

Finally, we remark that von Neumann considered the frequency interpretation of probability by von Mises [247–249] as the most adequate to the Copenhagen interpretation of QM (a footnote in his book [250], see also [167]).

1.2 Kolmogorov Axiomatics of Probability Theory

1.2.1 *Events as sets and probability as measure on a family of sets representing events*

The crucial point of the Kolmogorov approach is the *representation of random events by subsets* of some basic set Ω. This set is considered as *sample space* - the collection of all possible realizations of some experiment.[7] Points of sample space are called *elementary events*.

The collection of subsets representing random events shall be sufficiently rich to allow set-theoretic operations such as the intersection, the union, and the difference of sets. However, at the same time it shall not be too extended. If a too extended system of subsets is selected to represent events, then it may contain "events" which cannot be interpreted in any reasonable way (cf. the discussion on verification in Chapter 3).

After selection of a proper system of sets for events representation, we shall assign weights to these subsets:

$$A \mapsto P(A). \qquad (1.1)$$

The probabilistic weights are chosen to be *nonnegative real numbers and normalized to sum up to 1*: $P(\Omega) = 1$, the probability that anything indeed happens equals one.

---

[7]Consider the experiment of coin being tossed n times. Each realization of this experiment generates a vector $\omega = \{x_1, ..., x_n\}$, where $x_j = H$ (heads) or T (tails). Thus the sample space of this experiment contains 2^n points. We remark that this (commonly used) sample space is based on possible outputs of the observable corresponding to coin's sides and not on, so to say, hidden parameters of the coin and the hand leading to these outputs. Later we shall discuss this problem in more detail.

Interpretation: *An event with large weight is more probable (it occurs more often) than an event with small weight.*

We now discuss another feature of the probabilistic weights: the weight of an event A that can be represented as the disjoint union of events A_1 and A_2 is equal to the sum of weights of these events. The latter property is called *additivity* :

$$P(A_1 \cup A_2) = P(A_1) + P(A_2), \ A_1 \cap A_2 = \emptyset. \tag{1.2}$$

There is evident similarity with properties of mass, area, and volume.

It is useful to impose some restrictions on the system of sets representing events:

- (a1) set Ω containing all possible events and the empty set $\emptyset$ are also events (something happens and nothing happens);
- (a2) the union of two sets representing events represents an event;
- (a3) the intersection of two sets representing events represents an event;
- (a4) the complement of a set representing an event, i.e., the collection of all points that do not belong to this set, again represents an event.

Definition 1. *A set-system with properties (a1)-(a4) is called an algebra of sets (in the American literature, a field of sets).*

These set-theoretic operations correspond to the basic operations of classical *Boolean logic*: $\neg$ is the negation operator (NOT), $\wedge$ is the conjunction operator (AND), and $\vee$ is the disjunction operator (OR). The modern set-theoretic representation of events is a mapping of propositions describing events onto sets with preservation of the logical structure. The corresponding set-theoretic operations are denoted by the symbols $\complement$ (complement), $\cap$ (intersection), $\cup$ (union).[8] We recall that at the beginning of the mathematical formalization of probability theory the map (1.1) was defined on an algebraic structure corresponding to the logical structure, the *Boolean algebra* (invented by J. Boole [41], [42] the creator of Boolean logic).[9]

---

[8] We recall that $\complement A = \{\omega \in \Omega : \omega \notin A\}$. It is also convenient to use the operation of the difference of two sets: $A \setminus B = \{\omega \in A : \omega \notin B\}$, i.e., $A \setminus B = A \cap \complement B$.

[9] J. Boole tried to create a mathematical model of laws of mind [41]. In this way he created the Boolean logic and probability theory. Obviously, the laws of human mind and probability were indivisibly interconnected to him. Thus, the first lesson for students is that "classical probability" (at least in the set-theoretic framework) is fundamentally based on classical (Boolean) logic. Moreover, departures from classical probability may lead to departures from classical logic and vice versa.

In probabilistic considerations an important role is played by *De Morgan's laws*:

- *The negation of a conjunction is the disjunction of the negations.*
- *The negation of a disjunction is the conjunction of the negations.*

The rules can be expressed in formal language with two propositions A and B as:

$$\neg(P \wedge Q) \iff (\neg P) \vee (\neg Q), \quad \neg(P \vee Q) \iff (\neg P) \wedge (\neg Q). \qquad (1.3)$$

In the set-theoretic representation, De Morgan's laws have the form:

$$\complement(A \cup B) = \complement A \cap \complement B, \quad \complement(A \cap B) = \complement A \cup \complement B. \qquad (1.4)$$

From the set-theoretic representation of De Morgan's laws we see that in the definition of an algebra of sets it is possible to use only one of the conditions, (a2) or (a3).

Definition 2. *Let F be an algebra of sets. An additive map $\mu : F \to [0, +\infty)$ is called a measure.*

1.2.2   *The role of countable-additivity (σ-additivity)*

In the case of finite Ω the map given by (1.1) with the above-mentioned properties gives the simplest example of Kolmogorov's measure-theoretic probability. (Since Ω can contain billions of points, this model is useful for a huge class of applications.) Here $\Omega = \{\omega_1, ..., \omega_N\}$. To determine any map (1.1), it is enough to assign to each point $\omega \in \Omega$ its weight

$$0 \leq P(\omega_j) \leq 1, \ \sum_j P(\omega_j) = 1.$$

Then, by additivity this map is extended to the set-algebra consisting of all subsets of Ω :

$$P(A) = \sum_{\{\omega_j \in A\}} P(\omega_j).$$

However, if Ω is countable, i.e., it is infinite and its points can be enumerated, or "continuous" – e.g., a segment of the real line $\mathbf{R}$, then simple additivity is not sufficient to create a fruitful mathematical model. The map (1.1) has to be additive with respect to countable unions of disjoint events (σ-additivity)[10]:

$$P(A_1 \cup ... \cup A_n \cup ...) = P(A_1) + ... + P(A_n) + ..., \qquad (1.5)$$

[10]Here σ is a symbol for "countably".

and to work fruitfully with such maps (e.g., perform integration), one has to impose special restrictions on the system of sets representing events.

We need not just a set-algebra, but a σ-*algebra of sets* (in the American literature, a σ-field).

Definition 1σ. *An algebra of sets such that the generalizations of conditions (a2) and (a3) hold for countable unions and intersections of sets is called a σ-algebra of sets.*

In the logical terms, it means that, *new events (propositions) can be formed by applying the operations OR and AND countably many times.* We remark that in set theory De Morgan's laws hold for any system of sets (A_i), where $i \in I$ and I is some set of indices (maybe even uncountable). Therefore in the definition of a σ-algebra we can assume either closeness with respect to countable unions or countable intersections.

Definition 2σ. *Let $\mathcal{F}$ be a σ-algebra of sets.*

- σ-*additive map $\mu : \mathcal{F} \to [0, +\infty)$, i.e., with the condition of equality (1.5) holding for sequences of disjoint sets belonging $\mathcal{F}$, is called a σ-additive measure;*
- *a probability measure P is a σ-additive measure normalized by 1, $P(\Omega) = 1$.*

Thus by definition any probability measure is σ-additive.

Remark 1. Of course, this is a mathematical idealization of the real situation. Kolmogorov pointed out [177–179] that, since in real experiments it is impossible to "produce" infinitely many events, this condition is not experimentally testable. One may prefer to proceed with finite-additive probabilities. However, without σ-additivity it is difficult (although possible) to define the integral with respect to a probability measure (the Lebesgue integral is well defined only for σ-additive measures) and, hence, to define the mathematical expectation, the operation of averaging. J. L. Doob, another big name in the measure-theoretic approach to probability, also pointed out that "here may well be real world contexts for which the appropriate mathematical model is based on finitely but not countably additive set functions", p. 162 [80]. A similar problem arises in other mathematical models of probability, in particular, in the frequency approach to probability due to von Mises, see section 3.3.1 for the discussion.

One of the most important "continuous probability models" is based on the sample space $\Omega = \mathbf{R}$, i.e., elementary events are represented by real numbers, see section 3.1, for details.

1.2.3   *Probability space*

Let Ω be any set. Consider a σ-algebra of its subsets $\mathcal{F}$ and a probability measure P on $\mathcal{F}$. The Kolmogorov axiomatics (1933) [177] of probability theory can be presented in the following compact form:

Probability space is a triple $\mathcal{P} = (\Omega, \mathcal{F}, P)$.

Points ω of Ω are said to be *elementary events*, elements of $\mathcal{F}$ are *random events*, P is *probability*.

Remark 2. (Elementary events and random events). The terminology invented by Kolmogorov is a bit misleading. In fact, one has to distinguish between elementary and random events. The crucial point is that, in general, a single point set $A_\omega = \{\omega\}$, where ω is one of the points of Ω, must not belong to the σ-algebra of events $\mathcal{F}$. In this case we cannot assign the probability value to A_ω. Thus, some elementary events are, so to say, hidden; although they are present mathematically at the set-theoretical level, we cannot assign to them probability-values. One can consider the presence of such hidden elementary events as a classical analog of hidden variables in QM, although the analogy is not complete.

Example 1. Let us consider the sample space $\Omega = \{\omega_1, \omega_2, \omega_3\}$ and let the collection (algebra) of events $\mathcal{F}$ consist of the following subsets of $\Omega: \{\emptyset, \Omega, A = \{\omega_1, \omega_2\}, B = \{\omega_3\}\}$, and $P(A) = P(B) = 1/2$. Here the elementary single point events $\{\omega_1\}, \{\omega_2\}$ are not "physically approachable". (Of course, this is an indirect analogy.) Interpretations of the existence of "hidden elementary events" were not discussed so much in classical probability. There exist realizations of the experiment (represented by the probability space $\mathcal{P}$ with this property) such that it is in principle impossible to assign probabilities to them, even zero probability. We shall come back to this problem by discussing the classical probabilistic representation of observables.

1.3   Elementary Properties of Probability Measure

For beginners, we present some elementary properties of a probability measure. However, since the final output will be the famous Bell inequality, it is useful for everybody to go through the coming list of these properties.

1.3.1 *Consequences of finite-additivity*

In all formulas of this subsection, besides (1.12), the normalization by 1, i.e., $P(\Omega) = 1$, is not used. Here we explore only one basic feature of probability - its additivity. Below we shall present some elementary consequences of finite-additivity. An important consequence of σ-additivity of probability, its monotonicity, will be presented in section 1.3.3, see Chapter 3 for fundamental implications of σ-additivity.

Finite-additivity implies the following properties:

$$P(\emptyset) = 0, \tag{1.6}$$

i.e., the probability zero is assigned to the event of absence of any elementary event.

Proof. Since $\emptyset = \emptyset \cup \emptyset$ and $\emptyset \cap \emptyset = \emptyset$, we have $P(\emptyset) = P(\emptyset \cup \emptyset) = P(\emptyset) + P(\emptyset)$, and hence (1.6) takes place.

For any pair of events $A, B \in \mathcal{F}$,

$$P(A \cup B) = P(A) + P(B) - P(A \cap B). \tag{1.7}$$

This property generalizes the additivity formula for the disjoint events.

Proof. We use the following decompositions: $A \cup B = (A \setminus B) \cup (B \setminus A) \cup (A \cap B), A = (A \setminus B) \cup (A \cap B), B = (B \setminus A) \cup (A \cap B)$. Then LHS of (1.7) can be represented as $P(A \setminus B) + P(B \setminus A) + P(A \cap B)$, and RHS as $P(A \setminus B) + P(A \cap B) + P(B \setminus A) + P(A \cap B) - P(A \cap B)$. Hence, LHS=RHS.

For any event $B \in \mathcal{F}$, its complement $\complement B = \Omega \setminus B$ also belongs to $\mathcal{F}$. Let $A, B \in \mathcal{F}$, then

$$P(A) = P(A \cap B) + P(A \cap \complement B). \tag{1.8}$$

This trivial representation of the probability $P(A)$ has very important implications; for example, the formula of total probability and Bell's inequality.

Proof. Here we use the representation $A = (A \cap B) \cup (A \cap \complement B)$.

Additivity of probability in combination with its non-negativity implies that for any pair of events $A, B \in \mathcal{F}, A \subset B$,

$$P(A) \leq P(B). \tag{1.9}$$

Proof. Since $B = A \cup (B \setminus A)$, $P(B) = P(A) + P(B \setminus A)$. Non-negativity of probability implies (1.9).

From (1.7) we immediately obtain that

$$P(A \cup B) \leq P(A) + P(B) \qquad (1.10)$$

or more generally

$$P(\cup_i A) \leq \sum_i P(A_i), \qquad (1.11)$$

where $A_i \in \mathcal{F}$. Since P is not only finite-additive, but σ-additive, inequality (1.11) takes place for any countable union of events.

Finally, we remark that, since $P(\Omega) = 1$, we have

$$P(\complement A) = 1 - P(A). \qquad (1.12)$$

Exercise 1. Show that, for $A, B \in \mathcal{F}$, the following inequality holds:

$$|P(A) - P(B)| \leq P(A \Delta B), \qquad (1.13)$$

where the symmetric difference of two events is defined as $A \Delta B = (A \cup B) \setminus (A \cap B) = (A \setminus B) \cup (B \setminus A)$.

Exercise 2. Generalize equality (1.7) to the case of three sets, and then to the case of n sets, $n = 2, 3, \dots$.

We stress that in the above considerations we heavily used the Boolean structure of the system of events, i.e., that mathematically this system is represented by a set σ-algebra. Such a set-theoretic representation of events (together with additivity and non-negativity of probability) is very natural. It may seem that it is given from God. However, this is the idea of a human — Kolmogorov.

Of course, nobody can guarantee that all possible random experiments are covered by his axiomatics. The derived features of probability and their consequences (see our further considerations) put constraints on the domain of applicability of the Kolmogorov probability model. If, for some random experiment, one of these constraints is not satisfied, this simply means that a new model of probability has to be developed in order to match experiments of this type.[11] In such a situation one can speak about *non-Kolmogorovean probability*, cf. [2] - [4], [133], [156], [163].

---

[11]As was pointed out in the preface, R. Feynman was the first to present this viewpoint on inter-relation of classical probability and quantum physics [88]. He pointed out that probabilistic data collected in the two-slit experiment violates equality (1.8). He interpreted quantum interference of probabilities as a perturbation of this equality. Feynman's argument was reformulated in terms of conditional probabilities in a series of my works, e.g., [141], [142], [152], [161]. In fact, equality (1.8) is equivalent to the basic law of classical probability, the formula of total probability (FTP), section 1.6. Therefore in the probabilistic terms quantum interference is nothing else but a violation of FTP, see Chapter 5.

Now we derive the most famous constraint for classical probabilities widely known as Bell's inequality [31].

1.3.2   *Bell's inequality in Wigner's form*

Consider three events A, B, C. It is convenient to use the notations

$$A \equiv A_+, \complement A \equiv A_-, B \equiv B_+, \complement B \equiv B_-, C \equiv C_+, \complement C \equiv C_-.$$

Theorem 1. (Bell-Wigner inequality) *For any triple of events A, B, C, the following inequality holds:*

$$P(A_+ \cap B_+) + P(B_- \cap C_+) \geq P(A_+ \cap C_+). \tag{1.14}$$

Proof. For each term of (1.14), we shall use equality (1.8) which importance has already been emphasized. For the first term, the event $A_+ \cap B_+$ plays the role of A in (1.8) and the event C_+ plays the role of B in (1.8). We have

$$P(A_+ \cap B_+) = P(A_+ \cap B_+ \cap C_+) + P(A_+ \cap B_+ \cap C_-). \tag{1.15}$$

In the same way we obtain

$$P(B_- \cap C_+) = P(B_- \cap C_+ \cap A_+) + P(B_- \cap C_+ \cap A_-), \tag{1.16}$$

$$P(A_+ \cap C_+) = P(A_+ \cap C_+ \cap B_+) + P(A_+ \cap C_+ \cap B_-). \tag{1.17}$$

By adding the first two equalities we come to the expression

$$P(A_+ \cap B_+ \cap C_+) + P(A_+ \cap B_+ \cap C_-) + P(B_- \cap C_+ \cap A_+) + P(B_- \cap C_+ \cap A_-).$$

Commutativity of the operation of intersection implies that $P(A_+ \cap B_+) + P(B_- \cap C_+)$ equals to $P(A_+ \cap C_+)$ plus a nonnegative term. Hence, (1.14) holds.

We remark that to derive (1.14) we used only finite-additivity of probability. In Chapter 8 we shall present another form of the Bell type inequalities and discuss the physical interpretation of its violation in experiments with quantum systems.

1.3.3 *Monotonicity of probability*

Let $\mathcal{P} = (\Omega, \mathcal{F}, P)$ be a probability space.

Theorem 2. *For any monotonically decreasing sequence of events* $A_1 \supset A_2 \supset ... \supset A_n \supset A_{n+1} \supset ...$,

$$P(A) = \lim_{n \to \infty} P(A_n), \ A = \cap_{n=1}^{\infty} A_n. \tag{1.18}$$

Proof. We can always assume that $A = \emptyset$, otherwise we can operate with the sequence $A'_n = A_n \setminus A$. We have

$$A_1 = (A_1 \setminus A_2) \cup (A_2 \setminus A_3) \cup ...., ..., A_n = (A_n \setminus A_{n+1}) \cup (A_{n+1} \setminus A_{n+2}) \cup ....$$

Therefore the σ-additivity of P implies that

$$P(A_1) = \sum_{i=1}^{\infty} P(A_i \setminus A_{i+1}), ..., P(A_n) = \sum_{i=n}^{\infty} P(A_i \setminus A_{i+1}), ...$$

The series for $P(A_1)$ converges and the series for $P(A_n)$ is the remainder of the series for $P(A_1)$. The remainder of any convergent series goes to zero.

In the same way σ-additivity implies:

Theorem 3. *For any monotonically increasing sequence of events* $A_1 \subset A_2 \subset ... \subset A_n \subset A_{n+1} \subset ...$,

$$P(A) = \lim_{n \to \infty} P(A_n), \ A = \cup_{n=1}^{\infty} A_n. \tag{1.19}$$

In modern probability theory any consequence of σ-additivity of a probability measure is taken without any reservation, in spite of Kolmogorov's explicit statement that σ-additivity and its implications are ideal elements of theory and one has to be cautious in their practical applications, Remark 1, section 1.2.1. And it is clear why. Although Kolmogorov emphasized that σ-additivity is not testable experimentally, he chose this feature as the cornerstone of his theory, for reasons of mathematical simplicity and beauty. The position of R. von Mises, see section 3.3.1, is more consistent. He preferred to use less beautiful and more complicated mathematical formalism in which countable operations on events can lead to "events" to which probability cannot be assigned consistently.

1.4 Random Variables

In classical probability theory of Kolmogorov, random observations are represented by *random variables*. These are special maps from the space of elementary events to real numbers, $\xi : \Omega \to \mathbf{R}$ or to more general spaces which are typically endowed with some topological structures, e.g., the Euclidean space $\mathbf{R}^m$.[12] The specialty of maps representing random variables serves to establishing a fruitful integration theory with respect to corresponding probability measures in order to be able to determine numerical characteristics of random measurements, e.g., the average and the dispersion. Motivated by QM, in future we shall study in very detail mathematical models of random measurements, see Chapter 5. For the moment, we just stress that in classical probability theory, in its measure-theoretic model, random observations are mathematically represented by *maps*. We start with a brief introduction to theory of mathematically simplest random variables, the *discrete random variables*.

Definition 3d. *A discrete random variable on the Kolmogorov space* $\mathcal{P}$ *is a function* $a : \Omega \to X_a$, *where* $X_a = \{\alpha_1, ..., \alpha_n, ...\}$ *is a countable set (the range of values) such that all sets*

$$E^a_\alpha = \{\omega \in \Omega : a(\omega) = \alpha\}, \alpha \in X_a, \tag{1.20}$$

belong to $\mathcal{F}$.

Thus, as stressed above, classical probability theory is characterized by the *functional representation of observables*.

For each elementary event $\omega \in \Omega$, there is assigned a value of the observable a, i.e., $\alpha = a(\omega)$. This is the a-observation on realization ω of the experiment.

It is typically assumed that the range of values X_a is a subset of the real line. We will proceed under this assumption.

Remark 3. (Observations and hidden elementary events). Suppose the system of events $\mathcal{F}$ does not contain the single-point set for some $\omega_0 \in \Omega$. Thus the probability cannot be assigned to the ω_0-outcome of the random experiment. Nevertheless, this realization can happen and even the value of each observable is well defined: $a(\omega_0)$. However, we cannot "extract" information about this elementary event with help of the class of observables corresponding to the selected probability space. Consider the space from Example 1; let $\omega \to a(\omega)$ be a random variable, $\alpha_j = a(\omega_j)$. Suppose

---

[12]In many applications, e.g., in radio-engineering and telecommunication, optimization theory, random variables take values in infinite-dimensional spaces, e.g., in Hilbert or Banach spaces. Theory of such random variables is well developed. However, we do not plan to present it in this book. We mention it now because such infinite-dimensional random variables, with values in the space of squareintegrable functions L^2 (which is Hilbert space) will be formally used in sections 4.1, 5.2.2. There we use analogy to the theory of $\mathbf{R}$-valued random variables which we present in this section.

that $\alpha_1 \neq \alpha_2$. Then the sets $E^a_{\alpha_j} = \{\omega_j\}$. However, the single point sets $\{\omega_j\}, j = 1, 2$, do not belong to $\mathcal{F}$. Hence, we have to assume that $\alpha_1 = \alpha_2$ (otherwise subsets $E^a_{\alpha_j} = \{\omega_j\}$ should be measurable, since a is a random variable). And this is the general situation: we cannot distinguish such elementary events with the aid of observations.

The probability distribution of a (discrete) random variable a is defined as $p^a(\alpha) \equiv P(\omega \in \Omega : a(\omega) = \alpha)$. The *average* (mathematical expectation) of a random variable a is defined as

$$\bar{a} \equiv Ea = \alpha_1 \, p^a(\alpha_1) + ... + \alpha_n \, p^a(\alpha_n) + ... \qquad (1.21)$$

If the set of values of ξ is infinite, then the average is well defined if the series in the right-hand side of (1.21) converges absolutely.

Suppose now that the results of random measurement cannot be represented by a finite or countable set. This means that such an observable cannot be represented by a discrete random variable.

Definition 3. *A random variable on the Kolmogorov space $\mathcal{P}$ is any function $\xi : \Omega \to \mathbf{R}$ such that for any set Γ belonging to the Borel σ-algebra of the real line, its pre-image belongs to the σ-algebra of events $\mathcal{F}$:*

$$\xi^{-1}(\Gamma) \in \mathcal{F},$$

where $\xi^{-1}(\Gamma) = \{\omega \in \Omega : \xi(\omega) \in \Gamma\}$.

The space of random variables on the Kolmogorov probability space $\mathcal{P}$ is denoted by the symbol $R(\mathcal{P})$. This space will play an important role in formulation of conditions of compatibility of observables, in particular, quantum observables (see section 5.5).

In the non-discrete case the mathematical formalism is essentially more complicated. The main mathematical difficulty is to define the integral with respect to a probability measure on $\mathcal{F}$, the *Lebesgue integral* [177]. In fact, the classical probability theory, being based on the measure theory, is mathematically more complicated than the quantum probability theory. In the latter, integration is replaced by the trace operation for linear operators and the trace is always the discrete sum. Here we do not plan to present the theory of Lebesgue integration. We formally invent the symbol of integration.[13]

---

[13]We remark that, for a discrete random variable, the integral coincides with the sum for the mathematical expectation, see (1.21). Besides, a discrete random variable is integrable if its mathematical expectation is well defined. In general, any integrable random variable can be approximated by integrable discrete random variables, and the integral is defined as the limit of the integrals for the approximating sequence.

The *average* (mathematical expectation) of a random variable a is defined as

$$\bar{a} \equiv Ea = \int_{\Omega} \xi(\omega) dP(\omega). \tag{1.22}$$

The probability distribution of a random variable a is defined (for Borel subsets of the real line) as $p^a(\Gamma) \equiv P(\omega \in \Omega : a(\omega) \in \Gamma)$. This is a probability measure on the Borel σ-algebra. Calculations of the average can be reduced to integration with respect to p^a : $\bar{a} \equiv Ea = \int_{\mathbf{R}} x \, dp^a(x)$.

1.5 Conditional Probability; Independence; Repeatability

Kolmogorov's probability model is based on a probability space equipped with the operation of conditioning. In this model *conditional probability* is defined by the well-known *Bayes' formula*

$$P(B|C) = P(B \cap C)/P(C), P(C) > 0. \tag{1.23}$$

By Kolmogorov's interpretation it is the *probability of an event B to occur under the condition that an event C has occurred.* We emphasize that the Bayes formula is simply a definition, not a theorem; nor is it an axiom (this was particularly stressed by Kolmogorov); we shall see that in the frequency approach to probability the Bayes formula is a theorem.

We remark that conditional probability (for the fixed conditioning C)) $P_C(B) \equiv P(B|C)$ is again a probability measure on $\mathcal{F}$. For a set $C \in \mathcal{F}, P(C) > 0$, and a (discrete) random variable a, the conditional probability distribution is defined as $p_C^a(\alpha) \equiv P(a = \alpha|C)$. We naturally have $p_C^a(\alpha_1) + ... + p_C^a(\alpha_n) + ... = 1$, $p_C^a(\alpha_n) \geq 0$. The conditional expectation of a (discrete) random variable a is defined by $E(a|C) = \alpha_1 \, p_C^a(\alpha_1) + ... + \alpha_n \, p_C^a(\alpha_n) + ...$

Again by definition two events A and B are *independent* if

$$P(A \cap B) = P(A)P(B). \tag{1.24}$$

In the case of *nonzero probabilities* $P(A), P(B) > 0$ independence can be formulated in terms of conditional probability:

$$P(A|B) = P(A) \tag{1.25}$$

or equivalently

$$P(B|A) = P(B). \tag{1.26}$$

The *relation of independence is symmetric:* if A is independent of B, i.e., (1.25) holds, then B is independent of A, i.e., (1.26) holds, and vice versa.

(We remark that this property does not completely match our intuitive picture of independence.)

We now discuss an important feature of observables represented by random variables, namely, *repeatability*. We consider the following problem. Suppose that some discrete observable a was measured and the value $a = \alpha$ was registered. *What is the probability to obtain the same value $a = \alpha$ in successive measurement of the same observable a?*

In the Kolmogorov measure-theoretic framework we have (for discrete random variables):

$$P(a = \alpha | a = \alpha) = P(E_\alpha^a | E_\alpha^a) = P(E_\alpha^a \cap E_\alpha^a)/P(E_\alpha^a) = 1,$$

where $E_\alpha^a = \{\omega \in \Omega : a(\omega) = \alpha\}, \alpha \in X_a$. Here we used the idempotent feature of the Boolean operation of conjunction: $A \cap A = A$, for any set A. In the same way, for $i \neq j$, we have

$$P(a = \alpha_i | a = \alpha_j) = P(E_{\alpha_i}^a | E_{\alpha_j}^a) = P(E_{\alpha_i}^a \cap E_{\alpha_j}^a)/P(E_{\alpha_j}^a) = 0.$$

Here we used mutual disjointness of the sets E_α^a representing values of the observable a.

Thus *observables represented by random variables have the property of repeatability.*

To compare the conditions of repeatability in the Kolmogorov model and in the contextual probability model (section 5.3), it is convenient to write the conditions of repeatability in the following form:

$$P(a = \alpha_i | E_{\alpha_j}^a) = \delta_{ij}. \tag{1.27}$$

We remark that the property of repeatability for classical observables given by random variables is so natural (in the measure-theoretic framework) that typically it is even not discussed. In general its validity can be questioned and its violation sets bounds on applicability of the Kolmogorov model for description of random measurements, cf. with the case of quantum observables, sections 4.3 and 4.5.

1.6   Formula of Total Probability

In our further considerations the important role will be played by the *formula of total probability* (FTP). This is a theorem of the Kolmogorov model. Let us consider a countable family of disjoint sets A_k belonging to $\mathcal{F}$ such that their union is equal to Ω and $P(A_k) > 0, k = 1, \ldots$. Such a family is called a *partition* of the space Ω.

Theorem 4. *Let* $\{A_k\}$ *be a partition. Then, for every set* $B \in \mathcal{F}$, *the following formula of total probability holds*

$$P(B) = P(A_1)P(B|A_1) + ... + P(A_k)P(B|A_k) + ... \qquad (1.28)$$

Proof. We have

$$P(B) = P(B \cap (\cup_{k=1}^{\infty} A_k)) = \sum_{k=1}^{\infty} P(B \cap A_k) = \sum_{k=1}^{\infty} P(A_k) \frac{P(B \cap A_k)}{P(A_k)}.$$

Particularly interesting for us is the case when the partition is induced by a discrete random variable a taking values $\{\alpha_k\}$. Here,

$$A_k = E_{\alpha_k}^a \equiv \{\omega \in \Omega : a(\omega) = \alpha_k\}. \qquad (1.29)$$

Let b be another random variable. It takes values $\{\beta_j\}$. For any $\beta \in X_b$, we have

$$P(b = \beta) = \sum_k P(a = \alpha_k)P(b = \beta|a = \alpha_k). \qquad (1.30)$$

This is the basic formula of the Bayesian analysis. The probability to obtain the result β in the b-measurement can be estimated on the basis of conditional probabilities for this result (with respect to the results of the a-measurement).

1.7 Law of Large Numbers

Theorem 5. *(Kolmogorov Strong Law of Large Numbers [175], [177]) Let* $\xi_1, ..., \xi_N, ...,$ *be a sequence of identically distributed and independent random variables with average* $m = E\xi_j$. *Then*

$$P\left(\omega \in \Omega : \frac{\xi_1(\omega) + ... + \xi_N(\omega)}{N} \to m, N \to \infty\right) = 1, \qquad (1.31)$$

i.e.,

$$\lim_{N \to \infty} \frac{\xi_1(\omega) + ... + \xi_N(\omega)}{N} = m$$

almost everywhere (on the set of probability 1).

This law is the basic law guaranteeing the applicability of the Kolmogorov measure-theoretic model to experimental data: the arithmetic average approaches the measure-theoretic average (a.e.: for almost all sequences of trials). It was proven by Kolmogorov and this was the important step in justification of his model.

The main interpretation problem related to this law is of the following nature. Although the Kolmogorov strong law of large numbers implies that the set of elementary events for which the arithmetic mean converges to the probabilistic mean m is very large (from the measure-theoretic viewpoint), for the concrete ω (concrete sequence of experimental trials), this law cannot provide any information whether this convergence takes place or not. In fact, this was the main argument of von Mises (the creator of frequency probability theory) against the Kolmogorov measure-theoretic model, see the remark at the very end of section 1.10.

In principle, it might happen that the coin tossing would produce a sequence of a few hundred heads and no tails. In the discussion during my lectures in Vienna, Anton Zeilinger pointed out that one cannot exclude such a repeatability in say quantum experiments. And from the viewpoint of modern measure-theoretic approach to probability he is completely right. However, he had never seen such an experimental run and I am sure if it were produced, he would assign it not to mathematical features of zero probability, but to some problems in experiment preparations. I always had the feeling that there is something wrong with the conventional interpretation of the law of large numbers. I consider the frequency interpretation of measure-theoretic probabilities as one of the problems of the Kolmogorov model.[14]

Consider now some event A, i.e., $A \in \mathcal{F}$, and generate a sequence of independent tests in which the event A either happens or not. In the Kolmogorov model this sequence of experimental trials is described by a sequence of independent random variables, $\xi_j = 1$ if A happens and $\xi_j = 0$ in the opposite case. Then the frequency of the A-occurence, $\nu_N(A) = n(A)/N$, can be represented as

$$\nu_N(A; \omega) = \frac{\xi_1(\omega) + ... + \xi_N(\omega)}{N}. \tag{1.32}$$

Hence, the strong law of large numbers, in particular, implies that probabilities of events are approximated by relative frequencies:

$$P(A) = \lim_{N \to \infty} \nu_N(A; \omega), \tag{1.33}$$

but again with the same reservation: (1.33) takes place everywhere besides a set of measure zero. Thus in the Kolmogorov model the frequency interpretation of probability is derivable, it is a theorem. (As was remarked above, some probabilists, as von Mises, pointed out that the treatment of

---

[14]This problem is not present in the frequency approach to probability due to R. von Mises, section 1.10.

the strong law of large numbers as justifying the frequency interpretation has to be taken with caution.)

As was remarked, proving of the strong law of large numbers played the crucial role in coupling the Kolmogorov model with experiment. This version of the law of large numbers was an improvement of the law of large numbers which origin goes back to Bernoulli (Jacob) and Poisson with further contributions of Chebyshev, Markov, Borel, Cantelli, and with its final proof in the general measure-theoretic framework by Khinchin (therefore sometimes it is called Khinchin law of large numbers):

Theorem 6. (Law of Large Numbers) *Let* $\xi_1, ..., \xi_N, ...,$ *be a sequence of identically distributed and independent random variables with average* $m = E\xi_j$. *Then, for any* $\epsilon > 0$,

$$\lim_{N \to \infty} P\left(\omega \in \Omega : \left|\frac{\xi_1(\omega) + ... + \xi_N(\omega)}{N} - m\right| > \epsilon\right) = 0. \qquad (1.34)$$

Typically physicists are aware only about this probability convergence version of the law of large numbers, although heuristically they always use its strong version. However, the outputs of Theorems 5 and 6 differ essentially. By the latter we can only be sure that the arithmetic averages can deviate from the measure-theoretic average with a small probability. However, not only physicists, but even mathematicians (in 18th-19th centuries and at the beginning of 20th century) treated the latter as implying convergence of the arithmetic averages to the mean value. This is an important interpretation problem. In fact, to justify this "strong treatment" of the weak law of large numbers they used Cournot's principle for interpretation of probability, section 1.13 (see also the (b)-part of Kolmogorov's interpretation of probability). After Kolmogorov proved the strong law of large numbers, we need no more appeal to the probability convergence and apply the interpretational principles, as Cournot's principle, to justify the frequency interpretation of probability (see again section 1.13).

1.8 Kolmogorov's Interpretation of Probability

Kolmogorov proposed [177] to interpret probability as follows: "[. . .] we may assume that to an event A which may or may not occur under conditions Σ, [there] is assigned a real number $P(A)$ which has the following characteristics:

- (a) one can be practically certain that if the complex of conditions Σ is repeated a large number of times, N, then if n is the number of occurrences of event A, the ratio n/N will differ very slightly from $P(A)$;

- (b) if $P(A)$ is very small, one can be practically certain that when conditions Σ are realized only once the event A would not occur."

The (a)-part of this interpretation is nothing but the frequency interpretation of probability, cf. with von Mises theory and his principle of the statistical stabilization of relative frequencies, section 1.10. In the measure-theoretic approach this viewpoint on probability is justified by the *law of large numbers*. Of course, the referring to "practically certain" generates the problem which was discussed in section 1.7. However, for Kolmogorov, approximation of probability by frequencies was not the only characteristic feature of probability. The (b)-part (known in foundations of probability as *Cournot's principle*, section 1.13) also plays an important role. This is the purely weight-type argument: if the weight assigned to an event is very small then one can expect that such an event would never happen. We emphasize that Kolmogorov presented this weight-type argument in its strongest form - "never happen". One may proceed with a weaker form - "practically never happen", cf. section 1.13): weak and strong forms of the Cournot's principle.

1.9   Random Vectors; Existence of Joint Probability Distribution

In the Kolmogorov model a system of observables is represented by a vector of random variables $a = (a_1, ..., a_n)$. Its probability distribution is defined as

$$p_a(A_1 \times \cdots \times A_n) = P(\omega : a_1(\omega) \in A_1, \cdots, a_n(\omega) \in A_n). \qquad (1.35)$$

Thus, for any finite system of observables, it is assumed that the joint probability distribution exists.

Suppose that all coordinates of a random vector are discrete. Then $p_a(\alpha_1, ..., \alpha_n) = P(\omega : a_1(\omega) = \alpha_1, \cdots, a_n(\omega) = \alpha_n) = P(E_{\alpha_1} \cap ... \cap E_{\alpha_n})$. Thus to define the probability distribution, we have to appeal to Boolean logic.

1.9.1   *Marginal probability*

We remark that, for any subset of indices $i_1, ..., i_k$, the probability distribution of the vector $(a_{i_1}, ..., a_{i_k})$, $p_{a_{i_1}, ..., a_{i_k}}(A_{i_1} \times ... \times A_{i_k}) = P(\omega \in \Omega :$

$a_{i_1}(\omega) \in A_{i_1}, ..., a_{i_k}(\omega) \in A_{i_k})$, can be obtained from p_a as its marginal distribution. For example, consider a pair of discrete observables $a = (a_1, a_2)$. Then

$$p_{a_1}(\alpha_1) = \sum_{\alpha_2} p_a(\alpha_1, \alpha_2); \ p_{a_2}(\alpha_2) = \sum_{\alpha_1} p_a(\alpha_1, \alpha_2). \tag{1.36}$$

These conditions can be considered as conditions of *marginal consistency*. Consider now a triple of discrete observables $a = (a_1, a_2, a_3)$. Then

$$p_{a_1,a_2}(\alpha_1, \alpha_2) = \sum_{\alpha_3} p_a(\alpha_1, \alpha_2, \alpha_3), ..., p_{a_2,a_3}(\alpha_2, \alpha_3) = \sum_{\alpha_1} p_a(\alpha_1, \alpha_2, \alpha_3).$$

$$\tag{1.37}$$

One dimensional distributions can be obtained from the two dimensional distributions in the same way as in (4.12.1).

We remark that the possibility to represent two dimensional probability distributions as marginals of the joint probability distribution is the basic assumption for derivation of the Bell-type inequalities. In section 1.9.2 we shall discuss this point in more detail (see also Chapter 8).

1.9.2 *From Boole and Vorob'ev to Bell*

Theorem 1 (Bell-Wigner inequality) can be formulated by using random variables.

Theorem 1a. *Let* $a_i, i = 1, 2, 3$, *be dichotomous random variables, i.e.,* $a_i = \pm 1$, *which are determined on the same probability space. Then the following inequality holds:*

$$P(a_1 = +1, a_2 = +1) + P(a_2 = -1, a_3 = +1) \geq P(a_1 = +1, a_3 = +1). \tag{1.38}$$

In physics this inequality is typically written by using the probability distributions of random variables:

$$p_{a_1 a_2}(++) + p_{a_2 a_3}(-+) \geq p_{a_1 a_3}(++). \tag{1.39}$$

These notations shadow the key role of the existence of the common probability space. To enlighten this role even sharper, we write the inequality (1.38) by using the complete mathematical notations:

$$P(\omega \in \Omega : a_1(\omega) = +1, a_2(\omega) = +1) + P(\omega \in \Omega : a_2(\omega) = -1, a_3(\omega) = +1)$$

$$\geq P(\omega \in \Omega : a_1(\omega) = +1, a_3(\omega) = +1). \tag{1.40}$$

We remark that in the probability theory the problem of representation of a family of observables by random variables on the same probability space has long history. Already G. Boole [41], [42] considered the following problem. There are given three observables $a_1, a_2, a_3 = \pm 1$ and their pairwise probability distributions $p_{a_i a_j}$. Is it possible to find a discrete probability measure $p(\alpha_1, \alpha_2, \alpha_3)$ such that all $p_{a_i a_j}$ can be obtained as marginal probability distributions? Moreover, he showed that the inequality which nowadays is known as the Bell inequality is a *necessary condition* for the existence of such p. Therefore historically it may be correct to call this inequality as the *Boole-Bell inequality*. I even tried to proceed with this name in some of my publications, e.g., [155], but it seems that Bell's name has already been rigidly associated with this inequality.

For Boole, it was not self-evident that any family of observables can be described with the aid of a single probability space. He understood well that if data can be collected only for pairwise measurements, but it is impossible to jointly measure the triple, then it can happen that the "Boole-Bell inequality" is violated. For him, such a violation did not mean a violation of the laws of classical probability theory. From the very beginning, classical probability theory was developed as theory about *observable events* and the corresponding probabilities. Roughly speaking, probability distributions of "hidden variables" and counterfactuals were completely foreign for the creators of the probability theory.

A. N. Kolmogorov pointed out [178], section 2, that each complex of experimental physical conditions determines its own probability space. In the general setup he did not discuss the possibility to represent the data collected for different (may be incompatible) experimental conditions with the aid of a single probability space, cf. with Boole [41], [42]. Therefore we do not know Kolmogorov's opinion on this problem. However, he solved this problem positively for one very important case. This is the famous *Kolmogorov theorem* guaranteeing the existence of the probability distribution for any stochastic process, see section 1.9.4.

In 1962 the Soviet probability scientist N. N. Vorob'ev [251] presented a complete solution of the problem of the existence of joint probability distribution for any family of discrete random variables. He used his criteria for problems of game theory (the existence of a mixed strategy) and in random optimization problems (the existence of an optimal solution). However, the reaction of the Soviet probabilistic school (at that time it was one of the strongest in the world) to his attempt to go beyond the Kolmogorov axiomatics was very negative. His work [251] was completely forgotten. Only in 2005 it was "found" by K. Hess and W. Philipp and used in the Bell debate [110] - [113].

We remark that in the probabilistic community the general tendency was to try to find conditions for the existence of a single probability space. The Kolmogorov theorem about the existence of a probability space for a stochastic processes, Theorem 8, section 1.9.4, is one of the most known "existence theorems." In this situation the works in which non-existence was the main point were (consciously or unconsciously) ignored. In particular, even the aforementioned works of G. Boole were also completely forgotten. Around 2000 they were "found" by Itamar Pitowsky who used them in the Bell debate in quantum physics [29], [31], [156] (see Chapter 8 for details).

Now we formulate the following fundamental problem (see Chapters 5, 8 for further discussions):

Can quantum probabilistic data be described by the classical (Kolmogorov) probability model?

The (Boole-)Bell inequality is a necessary condition for the existence of such a description: if data can be described by a single Kolmogorov probability space, then Theorems 1 and 1a hold true and, for such data, this inequality has to be satisfied. As we know [31], this inequality is violated for the quantum probabilistic data. The latter is true both for quantum theory [31] and experiment [17], [254], [101], [59], [169], [109], [102] (see also Chapter 8). Therefore in general *quantum data cannot be embedded in a Kolmogorov model.*[15]

1.9.3 *No-signaling in quantum physics*

As we have seen, the (Boole-)Bell inequality can be treated as a statistical test of Kolmogorovness of data. Now we discuss another statistical test related to Bell's type experiments. We turn again to the condition of marginal selectivity. As was noted at the very end of section 1.9.1, this condition, see (1.37) is necessary for the existence of the joint probability distribution for three observables involved in Bell's considerations, see, e.g., Theorem 1 (1a).

It is also clear that even the condition of pairwise marginal consistency, see (4.12.1) has to be satisfied. In physical discussions on Bell's inequality they are known as *no-signaling conditions*. Here random variables a_1 and a_2 represent observables which are measured in two labs which are spatially

---

[15]In the light of our previous analysis this conclusion seems to be totally justified. However, the real situation is more complicated (see Chapters 5, 8, also [168]).

separated. The distance between labs is supposed to be so long that a signal about selection of measurement of a_2 in lab2 propagating with the velocity of light cannot approach lab1 at the instance of measurement of a_1 (see Chapter 8). The probability distribution of a_1 cannot depend on the decision of an experimenter which random variable is measured in lab2. If he selects another random variable, say a'_2, then the joint probability distribution has to reproduce the same probability distribution for a_1. Thus additionally to (4.12.1) we need to have:

$$p_{a_1}(\alpha_1) = \sum_{\alpha_2} p_{a_1,a'_2}(\alpha_1, \alpha'_2); \ p_{a_2}(\alpha_2) = \sum_{\alpha_1} p_{a_1,a'_2}(\alpha_1, \alpha'_2). \quad (1.41)$$

We remark [156] (see also Chapter 5) that a violation of the conditions of marginal consistency for pair observables is equivalent to a violation of FTP (or additivity of probability). Suppose that outputs of equations (1.41) and (4.12.1) are not equal (at least for one index j). Denote their left-hand sides as $q_{a_1}(\alpha_j)$ and $q'_{a_1}(\alpha_j)$ and preserve the notation $p_{a_1}(\alpha_j)$ for the probability distribution of a_1. (In physics this corresponds to performing measurement only in lab1.) If one of the equalities $q_{a_1}(\alpha_j) = q'_{a_1}(\alpha_j), j = 1, 2$, is violated, then at least one of the sums for the joint probability distributions, say $q_{a_1}(\alpha_j)$, should differ from $p_{a_1}(\alpha_j)$, i.e.,

$$\delta_j \equiv \delta_{a_1|a_2}(\alpha_j) = p_{a_1}(\alpha_j) - \sum_{\alpha_2} p_{a_1,a_2}(\alpha_1, \alpha_2) \neq 0. \quad (1.42)$$

To couple this inequality to violation of FTP, we, as usual, rewrite the sum of joint probabilities as

$$\sum_{\alpha_2} p_{a_1,a_2}(\alpha_1, \alpha_2) = \sum_{\alpha_2} P(a_2 = \alpha_2)P(a_1 = \alpha_1|a_2 = \alpha_2).$$

We emphasize that QM is consistent with no-signaling, calculations of the joint probability distributions of compatible observables represented by commuting operators (see Chapter 4) lead to no-signaling. Therefore in experimental tests on violation of Bell's inequality it is important to check not only its violation, but also that signaling is not statistically significant. Of course, for statistician the terminology "signaling" can be disapointing. This is just about checking marginal consistency. We remark that this important point was practically missed in experimenting with Bell's inequality. As we found (with G. Adenier who did the main part of data analysis) [6], for example, the data from the pioneer experiments of A. Aspect [17] and G. Weihs [254] exhibit strong signaling (see [256] for Weihs' explanation of such statistical behavior).

We raised this problem in the quantum community and put tremendous efforts to get new data to check for signaling.[16] In preprint [171] I was very critical with respect to common neglecting of the problem of statistical analysis of signaling. I really cannot understand: how can one be happy by violating Bell's inequality, even strongly, but on the signaling landscape? May be this preprint led to checks of no-signaling in the recent crucial experiments [109], [102] (but may be this was just a coincidence). There were reported that the hypothesis of no-signaling cannot be rejected with sufficiently high level of statistical significance, e.g., [102]. As the reader can see, in this book I handle only foundational problems of probability and randomness and not statistics at all. (The latter has real foundational troubles, even comparing with troubles of probability and randomness.) However, I have to point out that the problem of statistically significant of rejection of signaling is nontrivial. The main question is about the magnitude of the coefficients δ_j which has to be taken into account and checked for statistical significance? I believe experimenters that in real experiments these coefficients are non-negligibly positive. Thus first of all one has to set some ϵ and then check whether experimental data leads to rejection of the hypothesis $H_0 = \{\delta_j \geq \epsilon\}$. At least for me, selection of ϵ which would reflect adequately the experimental technicalities producing signaling is a big problem.

We remark that recently E. Dzhafarov proposed to handle the violations of signaling and Bell's inequalities consistently and roughly speaking to check the hypothesis that the violation of Bell's inequality is not a consequence of signaling.[17] He presented his approach in the form of new

---

[16]However, it was really impossible. Of course, experimenters claimed that their data is clean from signaling and they will send data to us, but we were not able to get data, in spite of numerous reminders. At the same time other experimenters told me that signaling is always present in data and they tried to explain the numerous technicalities which can generate "signaling". I became really mad of this situation and published arXiv-preprint [171] saying that experiments which data were not uploaded for the open access have no result. In fact, I did not expect any reaction from experimenters' side. However, immediately after the publication of my arXiv-post leading experimental groups started to promise to upload their data for open access after publication of the corresponding paper, e.g., [109], [102]. May be they took seriously my statement that only Weihs' experiment [254] can be considered as having the result, but may be the recent appearance of promises to put data to the web has no correlation with posting of preprint [171] - it might be just a random coincidence. We remind that up to the date of posting of the preprint [171] Weihs' experiment was the only one Bell's type experiment for which data could be uploaded from the web (for some period, but then it also disappeared - after publication [6]).

[17]Dzhafarov is a psychologist. He actively worked on the problem known as *marginal*

inequality modifying Bell's inequality, see, e.g., the work of Dzhafarov and Kujala [187].

This approach has some similarity with my general studies on interrelation between an "amount of violation of Kolmogorovness" and an "amount of violation of Bell's inequality". (A series of papers and books on this problem [137], [133] - [135], [156], [161] was totally ignored by the quantum community.) In Theorem 8 (8a) the existence of the common probability space $\mathcal{P} = (\Omega, \mathcal{F}, P)$ serving for all pairs of observations is crucial. Suppose now, as in my cited works, that there is a family of probability spaces, for simplicity only probability measures are different, i.e., instead of one P, there is a family (P_u). In probability theory we often explore different distances between probabilities, see, e.g. [233]. Take one of them ρ. Then $\gamma = \sup_{u,v} \rho(P_u, P_v)$ can be used as a measure of non-Kolmogorovness. There were derived modified Bell's inequalities with deformations expressed through γ. Thus in real experiments one has to check not the "genuine Bell inequality", but its deformations due to the impossibility to guarantee the same probability measure for any pair of measurements. Signaling is one of the signs of such impossibility - non-Kolmogorovness (cf., Chapter 5).

1.9.4 *Kolmogorov theorem about existence of stochastic processes*

The notion of a random vector is generalized to the notion of a *stochastic process*. Suppose that the set of indices is infinite; for example, $a_t, t \in [0, +\infty)$. Suppose that, for each finite set $(t_1...t_k)$, the vector $(a_{t_1}...a_{t_k})$ can be observed and its probability distribution $p_{t_1...t_k}$ is given. By selecting $\Omega_{t_1...t_k} = \mathbf{R}^k$, $P_{t_1...t_k} = p_{t_1...t_k}$, and $\mathcal{F}_k$ as the Borel σ-algebra for $\mathbf{R}^k$, we obtain the probability space $\mathcal{P}_{t_1...t_k}$ describing measurements at points $t_1...t_k$. At the beginning of 20th century the main mathematical question of probability theory was whether it is possible to find a single probability space $\mathcal{P} = (\Omega, \mathcal{F}, P)$ such that all a_t be represented as random variables on this space and all probability distributions $p_{t_1...t_k}$ are induced by the same P :

$$p_{t_1...t_k}(A_1 \times \cdots \times A_k) = P(\omega \in \Omega : a_{t_1}(\omega) \in A_1, ..., a_{t_n}(\omega) \in A_n). \quad (1.43)$$

Kolmogorov found natural conditions for the system of measures $p_{t_1...t_k}$ which guarantee the existence of such a probability space, see [177].

---

selectivity. This is, in fact, the problem of marginal (in)consistency in our terminology or (no-)signaling in quantum physics [83], [84] (see also Chapter 9).

Theorem 7. *Let a family of probability distributions* $(p_{t_1...t_k})$ *satisfy the following conditions:*

- *For any permutation* $s_1...s_k$ *of indices* $t_1...t_k$,

$$p_{t_1...t_k}(A_{t_1} \times ... \times A_{t_k}) = p_{s_1...s_k}(A_{s_1} \times ... \times A_{s_k}). \qquad (1.44)$$

- *For two sets of indices* $t_1...t_k$ *and* $r_1,..,r_m$,

$$p_{t_1...t_k r_1...r_m}(A_{t_1} \times ... \times A_{t_k} \times \mathbf{R} \times ... \times \mathbf{R}) = p_{t_1...t_k}(A_{t_1} \times ... \times A_{t_k}). \qquad (1.45)$$

Then there exist a probability space $\mathcal{P} = (\Omega, \mathcal{F}, P)$ *and a stochastic process* $a_t(\omega)$ *on it such that the equality (1.43) holds. The conditions (1.44) and (1.45) are also necessary for the existence of a stochastic process.*

During the next 80 years analysis of properties of (finite and infinite) families of random variables defined on one fixed probability space was the main activity in probability theory.

We remark that in Kolmogorov's construction of the probability space for a stochastic process the space of elementary events Ω is selected as the set of all trajectories $t \to \omega(t)$. The random variable a_t is defined as

$$a_t(\omega) = \omega(t). \qquad (1.46)$$

Construction of the probability measure P serving all finite random vectors is a mathematically sophisticated task going back to construction of the Wiener measure on the space of continuous functions.

For our considerations on inter-relation between classical and quantum probabilities, it is useful to remark that in order to describe an elementary event one has to monitor the whole trajectory of the process. One can even say that such an elementary event is determined by experimental data, measurement of the stochastic process in various instances of time. Thus the Kolmogorov construction of a stochastic process is based not on "hidden variables" (e.g., the initial conditions) determining in advance future results of measurements, but rather on the results themselves.

1.10 Frequency (von Mises) Theory of Probability

Von Mises (1919) theory was the first probability theory [247] - [249] based fundamentally on the principle of the statistical stabilization of frequencies. Although this principle was heuristically used from the very beginning of probabilistic studies, only von Mises tried to formalize it mathematically and to establish it as one of the basic principles of probability theory. His theory is based on the notion of a *collective* (*random sequence*).

Consider a random experiment S and denote by $L = \{\alpha_1, ..., \alpha_m\}$ the set of all possible results of this experiment[18]. The set L is said to be the label set, or the set of attributes of the experiment S. We consider only finite sets L. Let us consider N trials for this S and record the results, $x_j \in L$. This process generates a finite sample:

$$x = (x_1, ..., x_N\}, x_j \in L. \tag{1.47}$$

A *collective* is an infinite idealization of this finite sample:

$$x = (x_1, ..., x_N, ...\}, x_j \in L, \tag{1.48}$$

for which the following two von Mises principles are valid.

Principle 1 (statistical stabilization).

This is the principle of the statistical stabilization of relative frequencies of each attribute $\alpha \in L$ of the experiment S in the sequence (1.48). Take the frequencies

$$\nu_N(\alpha; x) = \frac{n_N(\alpha; x)}{N}$$

where $\nu_N(\alpha; x)$ is the number of appearance of the attribute α in the first N trials. By the principle of the statistical stabilization, *the frequency $\nu_N(\alpha; x)$ approaches a limit as N approaches infinity, for every label $\alpha \in L$.*

This limit

$$P_x(\alpha) = \lim_{N \to \infty} \nu_N(\alpha; x)$$

is called *the probability of the attribute α of the random experiment S.*

Sometimes (when the collective is fixed) this probability will be denoted simply as $P(\alpha)$ (although von Mises would be really angry; he always wrote that operation with abstract probabilistic symbols, i.e., having no relation to random experiment, is meaningless and may lead to paradoxic conclusions). We now cite von Mises [248]:

"We will say that a collective is a mass phenomenon or a repetitive event, or simply a long sequence of observations for which there are sufficient reasons to

---

[18]R. von Mises did not consider probability theory as a purely mathematical theory. He emphasized that this is a physical theory such as, e.g., hydrodynamics. Therefore his starting point is a physical experiment which belongs to physics and not to mathematics. He was criticized for mixing physics and mathematics. But he answered that there is no way to proceed with probability as a purely mathematical entity, cf. with remark of A. Zeilinger on the notion of randomness, section 2.4.

believe that the relative frequency of the observed attribute would tend to a fixed limit if the observations were infinitely continued. This limit will be called the probability of the attribute considered within the given collective".

Principle 2 (randomness).

Heuristically it is evident that we cannot consider, for example, the sequence

$$x = (0, 1, 0, 1, 0, 1, ...)$$

as the output of a random experiment. However, the principle of the statistical stabilization holds for x and $P_x(0) = P_x(1) = 1/2$. To consider sequences (1.48) as objects of probability theory, we have to put an additional constraint on them:

The limits of relative frequencies have to be stable with respect to a place selection (choice of a subsequence) in (1.48).

In particular, x does not satisfy this principle.[19]

However, this very natural notion (randomness) was the hidden bomb in the foundations of von Mises' theory. The main problem was to define a class of place selections which induces a fruitful theory. The main and very natural restriction which was set by von Mises is that a place selection in (1.48) cannot be based on the use of attributes of elements. For example, one cannot consider a subsequence of (1.48) constructed by choosing elements with the fixed label $\alpha \in L$. Von Mises defined a place selection in the following way [248], p. 9:

PS "a subsequence has been derived by a place selection if the decision to retain or reject the nth element of the original sequence depends on the number n and on label values $x_1, ..., x_{n-1}$ of the $n-1$ preceding elements, and not on the label value of the nth element or any following element".

Thus a place selection can be defined by a set of functions

$$f_1, f_2(x_1), f_3(x_1, x_2), f_4(x_1, x_2, x_3), ..., f_n(x_1, ..., x_{n-1}), ... \tag{1.49}$$

each function yielding the values 0 (rejecting the nth element) or 1 (retaining the nth element). Since any place selection has to produce from

---

[19] A. Zeilinger commented that, although this sequence has the deterministic structure, in principle it can be generated by some intrinsically random experiment S. May be the probability of such output is zero, but the experimenter would be ready to see such an event. This statement matches well the ideology of the Kolmogorov measure-theoretic approach, but not of the von Mises frequency approach. By the latter, if one obtained such a sequence, one has to question the randomness of the experiment.

an infinite input sequence also an infinite output sequence, it also has to satisfy the following restriction:

$$f_n(x_1, ..., x_{n-1}) = 1 \text{ for infinitely many } n. \tag{1.50}$$

Here are some examples of place selections:

- choose those x_n for which n is prime;
- choose those x_n which follow the word 01;
- toss a (different) coin; choose x_n if the nth toss yields heads.

The first two selection procedures are law-like, the third selection random. It is clear that all of these procedures are place selections: the value of x_n is not used in determining whether to choose x_n.

The principle of randomness ensures that no strategy using a place selection rule can select a subsequence with different odds (e.g., for gambling) than a sequence that is selected by flipping a fair coin. Hence, the principle can be called *the law of excluded gambling strategy*. We cite Feller [87], pp. 198, 199:

"The painful experience of many gamblers have taught us the lesson that no system of betting is successful in improving the gambler chances . . .The importance of this statement was first recognized by von Mises, who introduced the impossibility of a successful gambling system as a fundamental axiom."

Let $x = (x_j)$ be a collective (random sequence) with the label set $L = \{0, 1\}$, i.e., $x_j = 0, 1$. Given a place selection, see (1.49), (1.50), let n_1 be the least n such that $f_n(x_1, ..., x_{n-1}) = 1$, n_2 is the next such n, etc. Then by the principle of randomness:

$$\exists \lim_{N \to \infty} \frac{1}{N} \sum_{k=1}^{N} x_{n_k} = \lim_{N \to \infty} \frac{1}{N} \sum_{k=1}^{N} x_k = P_x(1). \tag{1.51}$$

Remark 4. The Kolmogorov measure-theoretic model is solely about probability; there is nothing about randomness. In the von Mises frequency model randomness and probability cannot be separated.

Remark 5. Later we shall present a variety of approaches to the notion of randomness. The von Mises approach can be characterized as *randomness as unpredictability* (no chance to beat the roulette).

The definition (**PS**) suffered from the absence of the mathematical proof of the existence of collectives. Von Mises reaction to the critique from mathematicians was not constructive (from the mathematicians viewpoint). He replied: go to a casino and you will get plenty of random sequences. We can compare this viewpoint with Zeilinger's proposal on physical randomness, section 2.4.

One can summarize the results of extended mathematical studies on the von Mises notion of randomness as follows:

If a class of place selections is over extended, then the notion of collective is over restricted (in fact, there are no sequences where probabilities are invariant with respect to all place selections). Nevertheless, by considering invariance of probabilities with respect to special classes of place selections only it is possible to construct a sufficiently rich set of such "restricted collectives".

And von Mises himself was completely satisfied with the latter operational solution of this problem. He proposed [249] to fix a class of place selections which depends on the physical problem and consider the sequences of attributes in which probabilities are invariant with respect to just this class of selections. That is, he again tried to move this problem outside mathematical framework.

The frequency theory of probability is not, in fact, calculus of probabilities, it is rather the calculus of collectives which generates the corresponding calculus of probabilities. We briefly discuss some of the basic operations for collectives, see [249], [133], [156] for detailed presentations. We remark from the very beginning that operations with collectives are more complicated than set theoretical operations. We consider only one operation.

Operation of mixing of collectives and the basic properties of probability. Let x be a collective with the (finite) label space L_x (here it is convenient to index label spaces by collectives) and let $E = \{\alpha_{j1}, ..., \alpha_{jk}\}$ be a subset of L_x. Sequence (1.48) is transformed into a new sequence x_E by the following rule (this is the operation called *mixing*). If $x_j \in E$, then at the jth place there is written 1; if $x_j \notin E$ then there is written 0. Thus the label set of x_E consists of two points, $L_{x_E} = \{0, 1\}$. This sequence has the property of the statistical stabilization for its labels. For example,

$$P_{x_E}(1) = \lim_{N \to \infty} \nu_N(1; x_E) = \lim_{N \to \infty} \nu_N(E; x) = \lim \sum_{n=1}^{k} \nu_N(\alpha_{jn}; x),$$

where $\nu_N(E; x) = \nu_N(1; x_E)$ is the relative frequency of 1 in x_E which coincides with the relative frequency of appearance of one of the labels from E in the original collective x. Thus

$$P_{x_E}(1) = \sum_{n=1}^{k} P_x(\alpha_{jn}). \qquad (1.52)$$

To obtain (1.52), we only used the fact that *the addition is a continuous operation on the field of real numbers*[20] **R**.

It is possible to show that the sequence x_E also satisfies the principle of randomness.[21] Hence this is a new collective. By this operation any collective x generates a probability distribution on the algebra $\mathcal{F}_{L_x}$ of all subsets of L_x (we state again that we consider only finite sets of attributes):

$$P(E)(\equiv P_x(E)) = P_{x_E}(1) = \sum_{n=1}^{k} P_x(\alpha_{jn}). \qquad (1.53)$$

We present the properties of this probability below.

- **P1** *Probability takes values in the segment $[0,1]$ of the real line.*
  Since $P(E) = \lim_{N\to\infty} \nu_N(E; x) = \lim_{N\to\infty} \frac{n_N(E;x)}{N}$, and $0 \le \nu_N(E; x) = \frac{n_N(E;x)}{N} \le 1$, then $P(E) \in [0,1]$.
- **P2** *Probability that something happens equals to 1.*
  This is a consequence of the fact that for $E = L_x$ the collective x_E does not contain zeros.
- **P3** *Additivity of probability.*
  Here we use representation (1.53) of probability. If E_1 and E_2 are two disjoint subsets of L_x, then, for $E = E_1 \cup E_2$,

$$P(E) = \sum_{\alpha\in E} P(\alpha) = \sum_{\alpha\in E_1} P(\alpha) + \sum_{\alpha\in E_2} P(\alpha) = P(E_1) + P(E_2).$$

Thus, in the von Mises theory, probability is also a probability measure. However, in contrast to the Kolmogorov theory, properties **P1 – P3** are *theorems*, not axioms. It is interesting that in his book [177] Kolmogorov referred to von Mises theory for justification of the properties **P1 – P3** for probability measure.

---

[20] At the moment this remark that properties of (frequency) probability have some coupling with inter-relation of the algebraic and topological structures on the real line can be considered simply as a trivial statement of purely mathematical nature. However, later we shall see that the result - *additivity of probability is a consequence of the fact that* **R** *is an additive topological group* - has a deep probabilistic meaning and it can lead to creation of a family of generalized frequency probability models a la von Mises.

[21] Here we do not discuss the problem of existence of collectives. Suppose that they exist or consider some special class of place selections – the restricted principle of randomness, see section 2.1.1.

We recall that in the Kolmogorov model the Bayesian expression of conditional probability is simply a definition. In the von Mises model it is a theorem (based on another operation for collectives, the so-called partition of collectives, see [247] - [249]). Thus, heuristically, the frequency theory is better justified. However, formally it is less rigorous.

We can say that the von Mises model was *the first operational model of probability.* In some sense it has even higher level of operationalization than the quantum model of probability. In the latter randomnesses carried by a system and a measurement device are still separated, one is represented by a quantum state (density operator) and another by an observer (Hermitian operator or POVM). Von Mises collective unifies these randomnesses.

Since the frequency definition of probability induces the probability measure on the algebra of all subsets of the set L_x of the attributes of a random experiment inducing the collective x, one may think that the Kolmogorov model can be considered simply as emergent from the von Mises model. And von Mises advertised this viewpoint in his books. Even the Kolmogorov referring to the von Mises model to justify the axioms of the measure-theoretic probability may induce such an impression. However, this is not as simple issue as it might seem. Kolmogorov's sample space Ω (endowed with the corresponding (σ)-algebra of random events) is not the same as von Mises' space of experiments attributes. For repeatable experiments, Kolmogorov's elementary events are, in fact, sequences of trials, so, roughly speaking, Kolmogorov's probability is defined on the set of all possible von Mises collectives. Of course, since the Kolmogorov model is abstract, one can consider even probabilities on the set of experiment's attributes, as von Mises did. However, the latter would not correspond to repeatable experiments in the Kolmogorov scheme.

The von Mises model can be easily extended to the case of countable sets of the experiment attributes, $L = (\alpha_1, ..., \alpha_n, ...)$. However, extension to "continuous sets of attributes", e.g., $L = \mathbf{R}$, is mathematically very difficult. It will not lead us to the notion of a measurable set used in the theory of the Lebesgue integral.

In the measure-theoretic model of Kolmogorov the (strong) law of large numbers is interpreted as providing the frequency interpretation of probability. Von Mises did not agree with such an interpretation. He claimed that, opposite to his principle of the statistical stabilization guaranteeing convergence of relative frequencies to the probability in each concrete collective (random sequence), the law of large numbers is a purely mathematical statement which has no direct relation to experiment. His main critical argument was that knowing something with probability one is totally meaningless if one needs to know something about the concrete random sequence.

Thus one can say that *in the development of probability we selected the pathway based on simpler mathematics and rejected the pathway based on our heuristic picture of probability and randomness.*

1.11   Subjective Interpretation of Probability

This section and sections 1.12, 1.13 are devoted to complicated interpretational issues. In principle, the reader can jump to Chapter 2. In short, he has to know just that from the subjectivist viewpoint probability $P(A)$ is the degree of personal belief in non/occurrence of the event A. Thus, as was stated by one of the fathers of subjective probability theory de Finetti, *"probability does not exist!"* (as objective entity). On the other hand, for those who plan to read Chapter 6 on interpretations of QM, reading of this section can be a good preliminary training, especially towards understanding of the information interpretation of QM and QBism. But again: if you feel tired of long discussions on interpretational issues, then jump to Chapter 2.

We have already presented two basic mathematical models of probability, von Mises' frequency model and Kolmogorov's measure-theoretic model. Nowadays the first one is practically forgotten and the second one is widely used in engineering, telecommunications, statistical physics, chemistry, biology, psychology, social science. However, the von Mises approach played the crucial role in development of theory of individual random sequences - randomness as unpredictability (impossibility of a successful gambling system), although at the final stage of the mathematical formalization of randomness it was suppressed by two other approaches - randomness as complexity (Kolmogorov) and as typicality (Martin-Löf), see Chapter 2.

Each scientific theory consists of two parts: a mathematical model and an interpretation of the mathematical entities. Both models (of von Mises and Kolmogorov) are endowed with the *statistical interpretation of probability* : if the same complex of experimental conditions (experimental context) is repeated a large number of times then the frequencies of outcomes approach the corresponding probabilities. One of the differences between the approaches of von Mises and Kolmogorov is that, for the first one the statistical interpretation is the essence of probability, but for the second one probability is initially defined as a measure and then one has to prove a *theorem*, the strong law of large numbers, to motivate the frequency interpretation. We remark that the definition of probability as a measure is heuristically equivalent to assigning weights to elementary events. Thus Kolmogorov's definition of probability is (again heuristically) based on the procedure of weighting of events. (Here Kolmogorov generalized Laplace's definition of probability which was based on assigning of *equal weights* to all elementary events.) However, Kolmogorov did not want to use this weight-like measure-theoretic definition of probability as its interpretation, with the aid of the law of large numbers he moved to the frequency interpretation coinciding with von Mises' "genuine frequency interpretation":

- (a) "one can be practically certain that if the complex of conditions Σ is repeated a large number of times, N, then if n is the number of occurrences of event A, the ratio n/N will differ very slightly from $P(A)$;"

We remark that in the process of creation of his measure-theoretic axiomatics of probability theory Kolmogorov was strongly influenced by von Mises, see [177] and also [133], [156]. Therefore it might be that the frequency interpretation was also motivated by von Mises theory. The main argument against using the weight-interpretation is that it loses its heuristic attraction for continuous sample spaces such as $\Omega = [a, b], \Omega = \mathbf{R}, \Omega = C([a, b])$ (the space of continuous functions on $[a, b]$ which is endowed with, e.g., Winer measure). Here, for non-discrete probability measure, the weight of each single elementary event is zero and it is impossible to jump from zero probability of elementary events to non-zero probability of a non-elementary event.

In any event the weight-like viewpoint to probability did not disappear completely from the Kolmogorov interpretation of probability and its trace can be found in the following statement (known as Cournot's principle, section 1.13):

- (b) "if $P(A)$ is very small, one can be practically certain that when conditions Σ are realized only once the event A would not occur."

This (b)-part of Kolmogorov's interpretation of probability is totally foreign to von Mises' genuine frequency ideology.

Now we make a point: in a scientific theory, the same mathematical model can have a variety of interpretations. In particular, it happened with Kolmogorov's measure theoretic-model of probability. Besides the commonly used statistical interpretation, probability measures can also be interpreted in the framework of *subjective probability theory*.

This interpretation was used by T. Bayes as the basis of his theory of probability inference, see then Ramsey [225], de Finetti [69], Savage [226], Bernardo and Smith [33]. Here the probability $P(A)$ represents an *agent's personal, subjective degrees of belief* in non/occurrence of the event A. In contrast to the statistically interpreted probability which is objective by its nature the subjective probability is by definition not objective, so to say, "it does not exist in nature" independently of an agent assigning probabilities to events. This viewpoint on probability is in direct conflict with von Mises' viewing of probability theory as a theory of natural phenomena, similar to, e.g., hydrodynamics. Kolmogorov and the majority of Soviet probabilists (see section 1.12) also took the active anti-subjectivist position. Although Kolmogorov treated probability theory as a mathematical theory (so his viewpoint on probability theory did not coincide with Mises' viewpoint), he also interpreted it as representing objective feature of repeatable phenomena, statistical stability of them.

At the same time, since the measure-theoretic definition of Kolmogorov probability is heuristically based on the weighting-like procedure for events, it seems that the subjective interpretation matches well the mathematical framework of Kolmogorov probability spaces. Instead of assigning to events objective weights (as Kolmogorov proposed to do), subjectivists assign to events personal weights, each agent assigns to an event A his own degree of belief. This personalization of probability contradicts not only to the views of von Mises, Kolmogorov and all their followers, but even the basic methodology of modern science. And, for example, de Finetti understood this well and emphasized this in his exciting and provocative essay [68]. He started with a citation of the important science methodological statements of Tilgher, see [68], p. 169 (in all following citations the italic font was inserted by me):

"Truth no longer lies in an imaginary equation of the spirit with what is outside it, and which, being outside it, could not possibly touch it and be apprehended; truth is in the very act of the thinking thought. The absolute is not outside our knowledge, to be sought in a realm of darkness and mystery; it is in our knowledge itself. *Thought is not a mirror in which a reality external to us is faithfully reflected*; it is simply a biological function, a means of orientation in life, of preserving and enriching it, of enabling and facilitating action, of taking account of reality and dominating it."

This viewpoint, thought is just a biological function and not reflection of the objective features of external reality, was shared by de Finetti and used by him to question the conventional ideology of modern science [68], p. 169:

"For those who share this point of view, which is also mine, but which I could not have framed better than with these incisive sentences of Tilgher's [...], *what value can science have?* In what spirit can we approach it? Certainly, we cannot accept determinism; we cannot accept the "existence", in that famous alleged realm of darkness and mystery, of immutable and necessary "laws" which rule the universe, and we cannot accept it as true simply because, in the light of our logic, it lacks all meaning. Naturally, then, science, understood as the discoverer of absolute truths, remains idle for lack of absolute truths. But *this doesn't lead to the destruction of science;* it only leads to a different conception of science. Nor does it lead to a "devaluation of science": there is no common unit of measurement for such disparate conceptions. Once the cold marble idol has fallen in pieces, the idol of perfect, eternal and universal science that we can only keep trying to know better, we see in its place, beside us, a living creature, the science which our thought freely creates. A living creature: flesh of our flesh, fruit of our torment, companion in our struggle and guide to the conquest.

Nature will not appear to it as a monstrous and incorrigibly exact clockwork mechanism where everything that happens is what must happen because it could not but happen, and where all is foreseeable if one knows how the mechanism works. To a living science nature will not be dead, but alive; and it will be like a friend about whom one can learn in sweet intimacy how to penetrate the soul and spirit, to know the tastes and inclinations, and to understand the character, impulses and abandonments. So *no science will permit us say: this fact will come about, it will be thus and so because it follows from a certain law, and that law is an absolute truth.* Still less will it lead us to conclude skeptically: the absolute truth does not exist, and so this fact might or might not come about, it may go like this or in a totally different way, I know nothing about it. What we can say is this: *I foresee that such a fact will come about, and that it will happen in such a way, because past experience and its scientific elaboration by human thought make this forecast seem reasonable to me.*

Here the essential difference lies in what the "why" applies to: I do not look for why THE FACT that I foresee will come about, but why I DO foresee that the fact will come about. It is no longer the facts that need causes; it is our thought that finds it convenient to imagine causal relations to explain, connect and foresee the facts. Only thus can science legitimate itself in the face of the obvious objection that our spirit can only think its thoughts, can only conceive its conceptions, can only reason its reasoning, and cannot encompass anything outside itself."

This statement contains such charge of energy that even one treating probability objectively cannot reject it without deep analysis. Of course, primarily de Finetti is right that in scientific prediction "I foresee that such a fact will come about, and that it will happen in such a way, because past experience and its scientific elaboration by human thought make this forecast seem reasonable to me." We have only our thought and even existence of objective reality is just one of its fruits.[22] For me the essential difference lies in the interpretations of "human thought": either as personalized or as collective. In the above citation from de Finetti, it seems that "human thought" has the meaning of thought of a kind of the *universal agent* doing scientific research. (The same can be said about "human experience" in Chopra's statement, see the previous footnote.) If we take subjective probability as the degree of belief of such a universal agent, then the dispute about objectivity or subjectivity of probability would be resolved peacefully. If de Finetti were not assumed the existence of objective reality ruled by natural laws, but just assumed the use of the scientific experience of the mankind, represented as the universal thinking agent, then von Mises or at least Kolmogorov might agree that such kind of subjective probability has the right for existence. This is my playing with the universal agent perspective on the subjective probability which can be compared with Gnedenko's statement, section 1.12: "Subjective probabilities, if necessary, can be made objective."

However, (and this is the main point) de Finetti strongly supports the personal viewpoint on subjective probability and, hence, "human thought" and, "past experience and its scientific elaboration".[23]

This *personal agent viewpoint* is unsympathetic for the majority of scientists, especially those exploring natural sciences. One of the main problems is that subjectivity of probability leads to *subjectivity of cause*. To

---

[22] I cannot miss the possibility to cite here also Deepak Chopra - controversial New-Age guru (his email from November 3, 2015) : "Scientific theories are made in consciousness, experiments are designed in consciousness, observations are made in consciousness. There is no science without consciousness. Consciousness created science. We objectify our human experience in consciousness and called that reality! There is no reality without consciousness. Atoms and molecules are words invented by humans for experiences in consciousness. Using experiences to explain experiences is tautology. Scientists need more humility and less hubris."

[23] It is clear why. And now the reader will understand why I did cite Chopra - because Chopra's attempt to escape personalization of human experience led to global consciousness, a kind of consciousness of universe. De Finetti definitely did not want to be in the global consciousness club.

illustrate this subjectivity of "causal relations", de Finetti presented the following provocative example [68], pp. 179-180:

"... the essence of the idea of cause escapes me entirely: it only shows itself when I pass from what is already known to predicting the unknown, when the factual data affect our state of mind, when from the easy science of hindsight we want to get a rule of action for the future. Suppose it to have been observed that many times, after an eclipse there is a war. Why don't I say that the eclipse is a cause of war, and why do superstitious people believe it? And why do we call them superstitious? In saying that the eclipse is not a cause of war I mean that, if tomorrow I see an eclipse, the outbreak of war will not therefore seem more likely to me than if an eclipse had not happened. One who says that the eclipse is a cause of war would mean that for him, on the contrary, after an eclipse he would see the threat of war as imminent. I call him superstitious because his state of mind is different from mine and *from that of the society to which I belong,* because it clashes with the conception of the world which is the innermost part of my imagination and of the imagination of my century. But if I want to strip away the part of my thought that is my own creation, if I want to distill from my opinions the objective part, i.e., the part that is purely logical or purely empirical, I will have to recognize that I have no reason to prefer my state of mind to that of a superstitious person except that I actually feel the state of mind which is mine, while that of a superstitious person repels me. The example I gave is an extreme case, and it might seem paradoxical. But there are infinitely many others where it would not surprise a contemporary if I said that I do not know how to tell whether or not a causal relation exists; there are infinitely many cases that daily give rise to such discussions. I expressed my opinion: *the concept of cause is subjective.*"

Here by rejecting the position of "eclipse causing believers" and selecting the position of those "from that of the society to which I belong" de Finetti might explore the universal agent perspective. However, he treated two aforementioned positions as equally acceptable.

The rejection of objectivity of cause by de Finetti can be compared with rejection of causality by von Neumann in his interpretation of QM, section 6.4. (Causality is rejected in all versions of the Copenhagen interpretation of QM. However, its probabilistic nature was discussed most clearly in von Neumann book [250].) However, in contrast to de Finetti, by rejecting causality von Neumann did not reject objectivity of probability. He used the statistical interpretation of probability in its genuine von Mises' frequency version. Von Neumann "saved" objectivity of probability in the absence of causality by inveting the concept of *irreducible quantum randomness*. Bohr and Pauli also interpreted probability statistically and, for them, it was definitely objective. This objectivity was based on objectivity of outputs of classical measurement devices. In contrast to von Neumann, they did not need irreducible quantum randomness.

However, in general de Finetti's denial of objectivity of cause had to be sympathetic for Copenhagenists. Therefore it is surprising that in QM nobody tried to proceed with the subjective probability interpretation. Only recently C. Fuchs supported by R. Schack proposed to use in QM subjective probability and personal agent's perspective (see section 6.7.3). This interpretation of QM is known as Quantum Bayesianism (QBism).

For the fathers of QM, both Copenhagenists (as Bohr, Pauli, Dirac, von Neumann) and anti-Copenhagenists (as Einstein, De Broglie), probability was objective and statistical. Why? Why not subjective? One of the reasons for this was that all physicists learned probability starting with classical statistical mechanics and the statistical interpretation was firmly incorporated in their mind. Some of them were unable to give up even causality, but not statistical nature of probability, However, it seems that the main reason was that de Finetti's views on probability and more generally on scientific theory were *too revolutionary* even for "quantum folk". The latter still wanted to have solid objective ground - in classical world, the world of macroscopic measurement devices. But de Finetti tried to teach us that even in this macro-world neither probability nor cause are objective, they have to be treated subjectively, person dependent. It seems that even Copenhagenists would not accept such a position. By following de Finetti consistently, they should reconsider not only physics of microworld, as was done in the process of creation of QM, but even physics of macroworld. It is interesting that even such a brave guy, C. Fuchs, by exploring the subjective interpretation of probability in quantum physics was not ready (yet?) to say that now Bayesianism has to be extended to classical statistical physics and thermodynamics.[24]

---

[24] My personal viewpoint on the subjective interpretation of probability is sufficiently complicated. As a student of the Department of Mechanics and Mathematics of Lomonosov Moscow State University, I was lucky to have a few lectures of Kolmogorov. Then he became too ill to continue, but lectures were given by his former student A. N. Shiryaev; in any event for us "subjective probability" was the swearword. Therefore by working on quantum foundations [133], [161] I always keep the statistical interpretation of probability. However, recently by working in applications of quantum probability to cognition, psychology, decision making (Chapter 9) I started to think that the subjective interpretation is adequate for modeling of decision making process by an individual agent. Thus I support Gendenko's statement (section 1.12): "I think every method has

1.12   Gnedenko's Viewpoint on Subjective Probability and Bayesian Inference

Here we present the viewpoint of B. V. Gnedenko[25] on the use of the statistical interpretation in the Bayesian inference. This citation is from the interview [235], B. V. Gnedenko was interviewed by N. Singpurvalla and R. Smith in 2006 (the italic font for some sentences was inserted by me):

Singpurvalla: Professor Gnedenko, you said that you met a lot of mathematicians in your life. Did you meet de Finetti?

Gnedenko: Yes, in Italy. I remember him with a great pleasure. We had a discussion about subjective probability.

Singpurvalla: And what do you think of it?

Gnedenko: *I think he was wrong.* Science must not be personalized. Science must be objective. Personality is very important for the development of science, but science itself must be objective. I had many discussions about this subject with de Finetti, and with Professor Savage. He was very learned and modest. I think de Finetti and Savage were most important in that area.

Smith: De Finetti' s books on the theory of probability were translated into English; were they translated into Russian?

Gnedenko: No.

Singpurvalla: Professor Gnedenko, since you have objection against subjective probability, you must have objection against Bayesian statistics also?

Gnedenko: *I think every method has its possibilities and its limitations.*

Singpurvalla: I agree.

Gnedenko: And the Bayes method has limitations. But in practice, dealing with serious problems, we should try all possible methods. I think the Bayes method has good possibilities and this method must be developed, but it is not the only one.

Singpurvalla: So you think that the Bayesian methods with objective probability are acceptable, but Bayesian methods with subjective probability are not?

Gnedenko: It is always necessary to state the assumptions. *Subjective probabilities, if necessary, can be made objective.*

Singpurwalla: Are there Bayesians in the Soviet Union? Belyaev has been writing Bayesian papers.

Gnedenko: Yes. Many Russian statisticians use Bayesian methods. It is impossible not to use them. These methods are important.

Singpurwalla: Have you used them?

Gnedenko: Yes. I have given my students Bayesian lectures.

Singpurwalla: I'd like to make a comment. When we started this interview, you said that your father wanted you to be a doctor, but you became a mathematician. Your son is a mathematician. Did you advise him to be mathematician?

---

its possibilities and its limitations."

[25] Gnedenko is known worldwide as the author of the textbook "Theory of Probability", see, e.g., [103], one of the best textbooks on probability. He contributed to theory of limit theorems, but also mathematical and applied statistics, including theory of reliability.

Gnedenko: He understands applications in psychology very well. It is impossible to do good work in mathematics without applications.

1.13 Cournot's Principle

Now after the brief introduction to subjective probability we once again turn to Kolmogorov's interpretation, section 1.8. Its (b)-part is also known as *Cournot's principle*. (In the presentation of this principle and historical circumstances of its appearance we follow the works of G. Shafer, see, e.g., [232].) Its first version is due to J. Bernoulli (1713) who related mathematical probability to *moral certainty/impossibility*:

"Something is morally certain if its probability is so close to certainty that the shortfall is imperceptible."

"Something is morally impossible if its probability is no more than the amount by which moral certainty falls short of complete certainty."

By setting the level of moral certainty (or impossibility) Bernoulli connected mathematical probability with the real world:

"Because it is only rarely possible to obtain full certainty, necessity and custom demand that what is merely morally certain be taken as certain. It would therefore be useful if fixed limits were set for moral certainty by the authority of the magistracy—if it were determined, that is to say, whether 99/100 certainty is sufficient or 999/1000 is required. . . "

In other words, *an event with a small probability will not happen!* A. Cournot (1843) stated that this principle of impossibility is the only way to connect mathematical probability with the real world (the name Cournot's principle was proposed by M. Fréchet). A. Cournot used this principle to justify the frequency interpretation of probability. At that time he already could refer the Bernoulli-Poisson version of the law of large numbers with probability convergence (which was later elaborated by Khinchin to Theorem 6, section 1.7). By using this principle he concluded that the event "the limit of the arithmetic averages would differ from the probability average" would never happen, since its probability is very small.

This was an important logical step in foundations of probability theory. In 18th-19th centuries, probabilists were not yet able to prove the strong law of large numbers, However, they as well as statisticians and physicists badly wanted to apply the frequency interpretation of probability. The use of Carnot's principle with combination of the weak law of large numbers "solved" this problem.

Cournot discussed not only moral impossibility (very small probability), but also *physical impossibility* (infinitely small probability):

"A physically impossible event is one whose probability is infinitely small."

In spite of the remark, section 1.8, that there is a subjective element in the (b)-part of Kolmogorov's interpretation (and Cournot's principle) - setting the level of moral impossibility, those who used this principle treated probability objectively. Subjectivists, as de Finetti, rejected it.

Cournot's principle principle played an important role in creation of objective interpretation of probability. At the beginning of 20th century it was supported by the majority of leading probabilists, e.g., by Hadamar, Lèvy, Borel, Kolmogorov, Frèchet. Frèchet distinguished between the weak and strong forms of Cournot's principle. The weak form says an event of small probability seldom happens. (Frèchet, Cramer). The strong form says an event of small probability will not happen (Cournot, Hadamard, Lèvy, Kolmogorov, Borel). As well as von Mises, Frèchet and Lèvy agreed that Cournot's principle leads to an objective concept of probability: Probability is a physical property just like length and weight.

However, in late 1950s and 1960s Cournot's principle practically disappeared from the probability scene. There were a few reasons for this. First of all, Kolmogorov after publication of his fundamental book on axiomatics of probability theory did not discuss the interpretational issues so much (may be not all). As was already pointed out, his model of probability started to be used "interpretationally free" - probability is just a measure. Another reason was strong attacks of subjectivists to the objective viewpoint on probability, especially de Finetti's critique. Moreover, as we have seen in Kolmogorov's interpretation, Cournot's principle is supplementary to the first part of this interpretation which is purely frequentist. And frequentists, as von Mises, tried to proceed just using the (a)-part of Kolmogorov's interpretation. They neither support Cournot's principle. Finally, after Kolmogorov proved the strong version of the law of large numbers, one need not more use this principle as a supplement to the weak version of this law (Bernoulli-Poisson-Khinchin). At the same time Kolmogorov by himself still explored Cournot's principle as the second part of his interpretation of probability and it was after he proved Theorem 5 [175]. And moreover the strong law of large numbers was presented in his monograph [177].[26]

---

[26]So, it seems a lot of historicity and psychology was involved in creation of Kolmogorov's theory of probability. I remark that in general his foundational monograph makes the impression that it was written "per one time", without deep reflections. It really might be the case. Kolmogorov never put his work on creation of the modern axiomatics of

By the 1970s, only Prokhorov (a former student of Kolmogorov) carries Kolmogorov's fame, expressing the principle in very special form: *only probabilities close to zero or one are meaningful.* (In fact, as student of Moscow state university, I heard this interpretational statement, but I treated it as total nonsense. For me probability was a measure and nothing more.)

Finally, we remark that in 1939 J. Ville showed [245] that Cournot's principle can be restated as a *principle of market efficiency*:

If you never bet more than you have, you will not get infinitely rich.

Later economists rediscovered this prinicple, but without relation to Cournot's principle. The original work of Ville [245] is famous as presenting a strong objection for von Mises notion of collective (Chapter 2). Its part related to market efficiency was completely neglected.

---

probability theory to the top of his scientific achievements. In private conversations he prioritized the contribution to theory of dynamical systems. He also told (again private communications) that at that time the idea about the measure-theoretic formalization of probability "was in the air". May be the latter is correct, but at the same time if we compare his work with works of his main competitors, say von Mises, Borel, Bernstein, Frèchet, we would be surprised by clearness and shortness of Kolmogorov's formalization comparing with longly and darkly writings of others. For example, I heard many times from top Soviet probabilists about Bernstein's axiomatization of probability, preceding Kolmogorov's axiomatization. Finally, by writing this book I read Bernstein's book [34]. It is a real mess... Bernstein tried to make a step towards an abstract presentation of probability theory, but his work heavily suffered of longly algebraic concretizations. Thus after its reading I appreciated Kolmogorov's contribution to foundations of classical probability theory even more.

Chapter 2

Randomness

In this chapter we shall briefly present basics of various mathematical models of classical randomness:

(1) randomness as unpredictability (von Mises-Wald-Church),
(2) randomness as typicality (Laplace-Ville-Martin-Löf),
(3) randomness as high algorithmic complexity (Kolmogorov, Solomonoff, Chaitin, Schnorr).

Each approach by itself is well formalized mathematically. The starting points of these approaches are very different, but surprisingly it happens that the last two viewpoints on randomness are equivalent, see Theorem 6 (Schnorr [227]), section 2.3. Moreover, such random sequences are also random in the sense of the first approach: they are von Mises-Wald-Church collectives, Corollary 1, section 2.3. However, inter-relation of the first approach with the other two is sufficiently complicated.

The main problem is that the Church formalization of the notion of a collective does not match completely the original viewpoint of von Mises. Church was the first who connected randomness with computability. He reduced the class of possible place selections to computable. In fact, there was no reason for this, besides mathematical simplification. Von Mises and Wald considered on the equal ground computable and non-computable place selections. Personally I am very critical to the commonly used Church approach to randomness based on computability. (We remark that the Martin-Löf and and Kolmogorov-Chaitin models of randomness were definitely stimulated by Church's attempt to justify the notion of collective by using computable place selections.) This approach matches well with mathematically constructible pseudo-random generators. However, there is no reason to proceed in this way by using physical sources of randomness.

And for them we do not have any mathematical theory of randomness.

In Chapter 7 we shall discuss quantum randomness and its inter-relation with classical randomness. This is merely a philosophic discussion. Its essence is that violation of causality of quantum measurements implies (heuristically) that their outputs form classically random sequences, in all aforementioned senses. At the same time it is clear that classical randomness, e.g., of coin tossing can be based on causal physical processes. Thus one may say that quantum random generators produce sequences which are "more random" than sequences produced by classical random generators. And this viewpoint is very popular in the quantum community. Surprisingly its justification depends heavily on an interpretation of QM. Acausality is the characteristic feature not of QM, but of the Copenhagen interpretation of QM (moreover, its special, although very popular, version due to von Neumann (see section 6.4). In section 5.5.3, we shall show that quantum probability can be treated as a very special calculus of classical conditional probabilities. Therefore outputs of groups of quantum measurements (as, e.g., in Bell's test) can be described by classical conditional probabilities. Thus the classical models of randomness can be used even for outputs of quantum measurements by taking into account conditioning on selection of experimental contexts. From this viewpoint, the aforementioned statements that quantum randomness differs very much from classical randomness seems to be not justified, see also section 8.4.

For me, the main problem is not inter-relation of classical and quantum randomnesses, but inter-relation between the commonly used mathematical ideology of randomness based on computability and "natural randomness" generated by physical, biological, social systems. The latter is definitely not computable and there is no mathematical theory of such randomness.

Now we start the presentation of the standard approaches to the notion of randomness based on the use of algorithmically realizable processes.

2.1 Random Sequences

At the heuristic level the difference between the notions of probability and randomness can be easily explained. First consider the set Ω of all sequences of the length N composed of zeros and ones, $x = (x_1, ..., x_N), x_j = 0, 1$. The uniform probability distribution on Ω is defined as $p(x) = 1/2^N, x \in \Omega$. Thus all sequences are equally probable. Suppose, for example, that in some experiment we obtained a sequence of the form (spaces are added to make it more readable):

110010010000111111011010101000100010000101101000110000100011010011

000100110001100110001010001011100000001101110000011100110100010 0

101001000000100100111000001000100010100110011111001100011 1010000

000010000010111011111101010011000111011000100111001101100 10001001

This sequence matches well our heuristics about randomness. To use it later, we denote this sequence x_{rand}. (Here $N = 258$.) Now suppose that we obtained the sequence of the same length of the form $x = 010101....01$ (composed of blocks 01). It is clear that it cannot be treated as a random sequence. Its appearance seems to be strange and unexpected for us (in a random experiment). However, the probability to pick up both sequences from Ω is the same. Thus the probability calculus cannot formalize our heuristic image of randomness. Of course, it is clear why the second sequence is not random. The repeatability of the appearance of zeros and onces is a definite signal of presence of some *causal process* producing this regular pattern. But in general it is very difficult to understand whether there is some causal process behind an observed sequence of zeros and ones. To illustrate this problem, I specially took the "random sequence" x_{rand} as the first 258 digits in the binary expansion of the number π which can be computed algorithmically. Thus one cannot simply proceed heuristically. Some formal theory of randomness and its inter-relation with probability has to be developed.

We remark that consideration of infinite sequences cannot solve the problem of mismatching heuristics related to probability and randomness. The space of infinite binary sequences Ω can also be endowed with the uniform probability distribution.[1] Sequences which look like random and sequences which look like regular do have the same probability of appearance (in fact, zero probability).

The problem of inter-relation of the notions of probability and randomness was discussed already by Laplace [70]:

"We arrange in our thought, all possible events in various classes; and we regard as extraordinary those classes which include a very small number. In the

---

[1] The simplest way to introduce this probability measure is to map binary sequences into real numbers from [0,1]; this map from Ω to [0, 1] can be used to back transport measures from the interval [0,1] to Ω; to obtain the uniform probability distribution on Ω, one takes the Lebesgue linear measure on [0,1] which is defined through its values on intervals: $\mu([a, b)) = b - a$.

game of heads and tails, if heads come up a hundred times in a row, then this appears to us extraordinary, because the almost infinite number of combinations that can arise in a hundred throws are divided in regular sequences, or those in which we observe a rule that is easy to grasp, and in irregular sequences, that are incomparably more numerous."

Roughly speaking, Laplace tried to shift the problem of randomness from individual sequences to classes (sets) of sequences. Although we cannot distinguish random and non-random sequences using probabilities of their individual appearance, we might try to estimate probability measures of some classes of sequences. The set of regular sequences has a small measure; the set of random sequences has an essentially larger measure. From this viewpoint, random sequences are typical and regular are atypical. This viewpoint to formalization of the notion of randomness led to definition of randomness as *typicality* (Laplace-Ville-Martin-Löf, see section 2.1.2).

Besides the typicality dimension of randomness Laplace also pointed to the aforementioned presence of a causal process as preventing a sequence to be random [70]:

"The regular combinations occur more rarely only because they are less numerous. If we seek a cause whenever we perceive symmetry, it is not that we regard the symmetrical event as less possible than the others, but, since this event ought to be the effect of a regular cause or that of chance, the first of these suppositions, is more probable than the second. On a table, we see letters arranged in this order: C o n s t a n t i n o p l e, and we judge that this arrangement is not the result of chance, not because it is less possible than others, for if this word were not employed in any language we would not suspect it came from any particular cause, but this word being in use among us, it is incomparably more probable that some person has thus arranged the aforesaid letters than this arrangement is due to chance."

This viewpoint led to formalization of the notion of an individual random sequence through formalization of the notion of causal generation of a sequence which culminated in the Kolmogorov algorithmic complexity approach to randomness, sections 2.2, 2.3, see A. Dasgupta [65] for details.

2.1.1 *Approach of von Mises: randomness as unpredictability*

Von Mises did not solve the problem of the existence of collectives. Immediately after his proposal on collectives, mathematicians started to ask whether such sequences exist at all.

Kamke's attack to von Mises theory

One of the first objections to von Mises approach was presented by E. Kamke, see, e.g., [156] and is well known as *Kamke's objection*. The principle of randomness of von Mises implies stability of limits of frequencies with respect to the set of *all possible place selections*. Kamke claimed that there are no sequences satisfying this principle. Here we reproduce his argument.

Let $L = \{0,1\}$ be the label set and let $x = (x_1, ..., x_n, ...), x_j \in L$, be a collective which induces the probability distribution $P : P(0) = P(1) = 1/2$. Now we consider the set $S_{\text{increasing}}$ of all strictly increasing sequences of natural numbers, its elements have the form $k = (n_1 < n_2 < n_m < ...)$, where $n_j, j = 1, 2, ...$ are natural numbers. This set can be formed independently of the collective x. However, among elements of $S_{\text{increasing}}$, we can find the strictly increasing sequence $\{n : x_n = 1\}$. This sequence defines a place selection which selects the subsequence $(1, 1, ..., 1, ...)$ from the sequence x. Hence, the sequence x (for which we originally supposed that it is a collective) is not a collective after all!

However, the mathematical structure of Kamke's argument was not completely convincing. He claimed to have shown that for every putative collection x there exists a place selection ϕ that changes the limits of frequencies. But the use of the existential quantifier here is classical (Platonistic). Indeed, it seems impossible to exhibit explicitly a procedure which satisfies von Mises' criterion (independence on the value x_n) and at the same time selects the subsequence $(1, 1, 1, ...)$ from the original sequence x. In any event Kamke's argument played an important role in understanding that to create the mathematically rigorous theory of collectives one has to restrict the class of place selections.[2]

Mises-Church collectives (random sequences)

The simplest way (at least from the mathematical viewpoint) is to proceed with special classes of *lawlike place selections*. In particular, A. Church [60] proposed to consider place selections (1.49), (1.50) in which the selection functions $f_n(x_1, ..., x_{n-1})$ are algorithmically computable. (We recall that f_n is used to select or not the nth element of a sequence

---

[2]The work of Kamke as well as other works devoted to the analysis of the von Mises principle of the place selection also played the fundamental role in setting the foundations of theory of algorithms and constructive mathematics. In the latter the arguments similar to the Kamke objection are not taken into account, because they are not based on constructive (algorithmically performable) proofs.

$x = (x_1, ..., x_n, ...)$.) It is important to note that the set of all Church-like place selections is countable, see, e.g., [156]. The existence of Church's collectives is a consequence of the general result of A. Wald [253] which we formulate now.

Mises-Wald collectives (random sequences)

Let $p = (p_j = P(\alpha_j))$ be a probability distribution on the label set $L = \{\alpha_1, ..., \alpha_m\}$. Denote the set of all possible sequences with elements from L by the symbol L^∞. Let ϕ be a place selection. For $x \in L^\infty$, the symbol ϕx is used to denote the subsequence of x obtained with the aid of this place selection.

Let U be some family of place selections. We set

$$X(U; p) = \{x \in L^\infty : \forall \phi \in U \lim_{N \to \infty} \nu_n(\alpha_j; \phi x) = p_j, j = 1, ..., m\},$$

where as usual $\nu_N(\alpha; y), \alpha \in L$, denotes the relative frequency of the appearance of the label α among the first N elements of the sequence $y \in L^\infty$.

Theorem 1. (Wald [253]) *For any countable set U of place selections and any probability distribution p on the label set L, the set of sequences $X(U; p)$ has the cardinality of the continuum.*

By Wald's theorem for any countable set of place selections U, the frequency theory of probability can be developed at the mathematical level of rigorousness. R. von Mises was completely satisfied by this situation (see [249]).

Ville's attack to von Mises theory

However, a new cloud appeared on the sky. This was the famous *Ville's objection* [245].

Theorem 2. (Ville) *Let $L = \{0, 1\}$ and let $U = \{\phi_n\}$ be a countable set of place selections. Then there exists $x \in L^\infty$ such that*

- *for all n,*

$$\lim_{N \to \infty} \sum_{j=1}^{N} (\phi_n x)_j = 1/2;$$

- *for all N,*

$$\sum_{j=1}^{N} (\phi_n x)_j \geq 1/2.$$

Such a sequence x of zeros and ones is a collective with respect to U, $x \in X(U; 1/2)$, but seems to be far too regular to be called random. At the same time from the measure-theoretic viewpoint the existence of such sequences is not a problem. The set of such sequences has the Lebesgue measure zero. (We recall that any sequence of zeros and ones can be identified with a real number from the segment $[0, 1]$.) Here we see the difference between the treatment of randomness as unpredictability (a la von Mises) and as typicality (see section 2.1.2 for the latter – theory of Martin-Löf).

2.1.2 Laplace-Ville-Martin-Löf: randomness as typicality

Ville used Theorem 2 to argue that collectives in the sense of von Mises and Wald do not necessarily satisfy all intuitively required properties of randomness. J. Ville introduced [245] a new way of characterizing random sequences (cf. with Laplace [70]), based on the following principle:

Ville's Principle: *A random sequence should satisfy all properties of probability one.*[3]

Each property is considered as determining a *test of randomness*. According to Ville [245], a sequence can be considered random if it passes all possible tests for randomness.

However, this is impossible: we have to choose countably many of those properties (otherwise the intersection of the uncountable family of sets of probability 1 can have probability less than 1 or simply be nonmeasurable; in the latter case the probability is not defined at all). Countable families of properties (tests for randomness) can be selected in various ways. A random sequence passing one countable sequence of tests can be rejected by another. This brings ambiguity in the Ville approach to randomness as typicality (i.e., holding some property with probability one).

It must be underlined that the Ville principle is really completely foreign to von Mises. For von Mises, a collective $x \in L^\infty$ induces a probability on the set of labels L, not on the set of all sequences L^∞. Hence, for von Mises (and other scientists interpreting randomness as unpredictability in an individual sequence), there is no connection between properties of probability one in L^∞ and properties of an individual collective.

Later (in 1970s) Per Martin-Löf (motivated by the talk of Andrei Nikolaevich Kolmogorov at the Moscow Probability Seminar) [198] solved the

---

[3] Of course, Ville's approach matches Laplace viewpoint [70] on randomness as typicality.

problem of ambiguity of the Ville interpretation of randomness as typicality. He proposed to consider *recursive (algorithmic) properties of probability one*, i.e., the properties which can be tested with the aid of algorithms. Such an approach induces the fruitful theory of recursive (algorithmic) tests for randomness (see, for example, [199], [200]). The key point of this "algorithmic tests" approach to the notion of randomness is that it is possible to prove that there exists the *universal algorithmic test*. A sequence is considered random if it passes this universal test. Thus the class of typicality-random sequences is well-defined. Unfortunately, this universal test of randomness cannot be constructed algorithmically (although by itself it is an algorithmic process). Therefore, although we have the well-defined class of Martin-Löf random sequences, we do not know what the universal test for such randomness looks like. Hence, for a concrete sequence of trials we cannot check algorithmically whether it is random or not – although we know that it is possible to perform such algorithmic check.

So, as in the case of Mises-Wald randomness (as unpredictability), Ville-Martin-Löf randomness (as typicality) cannot be considered satisfactory.

Finally, we remark that similar approach, randomness as recursive (algorithmic) typicality, was developed by Schnorr [227].[4]

2.2   Kolmogorov: Randomness as Complexity

It is well known that personally A. N. Kolmogorov was not satisfied by his own measure-theoretic approach to probability (private communications of his former students). He sympathized to the von Mises approach to probability in which randomness was no less fundamental than probability. In 1960s he turned again to the foundations of probability and randomness and tried to find foundations of randomness by reducing this notion to the notion of complexity, 1963 [180] - [183]. Thus in short the Kolmogorov approach can be characterized as *randomness as complexity*. Another basic point of his approach is that complexity of a sequence has to be checked *algorithmically*.

Let $L = \{0, 1\}$. Denote by L^* the set of all *finite sequences* (words) in the alphabet L.

Definition 1 (Kolmogorov). *Let A be an arbitrary algorithm. The*

---

[4]We state again that approaches of Martin-Löf and Schnorr (as well as Ville) have nothing to do with the justification of Mises' frequency probability theory and his viewpoint on randomness as unpredictability.

algorithmic complexity of a word x with respect to A is $K_A(x) = \min l(\pi)$, *where* $\{\pi\}$ *are the programs which are able to realize the word x with the aid of A.*

Here $l(\pi)$ denotes the length of a program π. This definition depends on the structure of the algorithm A. Later Kolmogorov proved the following fundamental theorem:

Theorem 3. (Kolmogorov, Solomonoff)[5] *There exists an algorithm* A_0 *(optimal algorithm) such that, for any algorithm A, there exists a constant* $C > 0$,

$$K_{A_0}(x) \leq K_A(x) + C. \tag{2.1}$$

It has to be pointed out that optimal algorithm is not unique.

The complexity $K(x)$ of the word x is (by definition) equal to the complexity $K_{A_0}(x)$ with respect to one fixed (for all considerations) optimal algorithm A_0.

The original idea of Kolmogorov [180] - [183] was that complexity $K(x_{1:n})$ of the initial segments $x_{1:n}$ of a random sequence x has to have the asymptotic $\sim n$

$$K(x_{1:n}) \sim n, n \to \infty, \tag{2.2}$$

i.e., we might not find a short code for $x_{1:n}$.

However, this heuristically attractive idea was rejected as a consequence of the objection of Martin-Löf [200].

To discuss this objection and connection of the Kolmogorov complexity-randomness with Martin-Löf typicality-randomness, we proceed with *conditional algorithmic complexity* $K(x; n)$, instead of complexity $K(x)$. Conditional complexity $K(x; n)$ is defined as the length of the minimal program π which produces the output x on the basis of information that the length of the output x is equal to n.

---

[5]Kolmogorov published this result in 1965 [134]. In fact, the first proof was presented in Solomonoff's preprint [237] in 1960. However, practically nobody paid attention to this work. Its importance was not recognized by the scientific community. When Kolmogorov became aware of this work, he openly recognized the priority of Solomonoff. In fact, this Kolmogorov's recognition played the crucial role in advertising research of Solomonoff who first became famous in Soviet Union and only later in Western countries. Kolmogorov's contribution to establishing this area of research was memorized in assigning the name *Kolmogorov complexity* to the algorithmic complexity. This does not diminish the role of the contribution of Solomonoff [237], [238] (and Chaitin [54]).

Theorem 4. (Martin-Löf) *For every binary sequence* x,

$$K(x_{1:n}; n) < n - \log_2 n, \qquad (2.3)$$

for infinitely many n.

Hence, *Kolmogorov random sequences, in the sense of the definition (2.2), do not exist.*

Another problem of the Kolmogorov approach to randomness as algorithmic complexity is that we "do not know" any optimal algorithm A_0, i.e., *the Kolmogorov complexity is not algorithmically computable!* However, the latter problem can be partially fixed, because there exist algorithmic methods to estimate this complexity (from above and from below).

Finally, we remark that *numerical evaluation of the Kolmogorov complexity for binary sequences produced in experiments with quantum systems, see Chapter 7, is an open and interesting problem.*

2.3   Kolmogorov-Chaitin Randomness

As we have seen, in its straightforward form the Kolmogorov algorithmic complexity does not lead to a nontrivial notion of a random sequence. However, as it happens [54], a fruitful theory of individual random sequences is very near, it is enough to slightly modify the notion of Kolmogorov complexity.

First, we remark that the notion of algorithm can be formalized as a *partial computable (recursive) function* $A : L^* \rightarrow L^*$ (here "partial" means that in general such a function is defined only on some subset $D = D_A$ of L^*). Thus Kolmogorov complexty of a word $x \in L^*$ with respect to A equals the length of the shortest word $\pi \in D$ such that $A(\pi) = x : K_A(x) = l(\pi)$. If such $\pi \in D$ does not exist, then $K_A(x) = +\infty$.

A prefix of a word $x = x_1 \ldots x_n$ is a $\widehat{x} = x_1 \ldots x_m$, where $m \leq n$. A subset D of L^* is called *prefix free* if no word in D is a prefix of another member of D. A real world example of a prefix free set (over the alphabet of decimal digits) is the set of country dialing codes in the international telephone system.

Definition 2. *Let A be an arbitrary algorithm, a partial computable function, with prefix free domain of definition D. Algorithmic prefix free complexity of a word $x \in L^*$ with respect to A equals to the length of the shortest word $\pi \in D$ such that $A(\pi) = x : \tilde{K}_A(x) = l(\pi)$. If such $\pi \in D$ does not exist, then $\tilde{K}_A(x) = +\infty$.*

Theorem 5. *There exists an optimal prefix free algorithm (partial computable function) A_0 such that, for any prefix free algorithm A, there exists a constant $C > 0$,*

$$\tilde{K}_{A_0}(x) \leq \tilde{K}_A(x) + C. \tag{2.4}$$

Prefix free complexity $\tilde{K}(x)$ of the word x is (by definition) equal to the complexity $K_{A_0}(x)$ with respect to one fixed (for all considerations) optimal prefix free algorithm A_0.

Definition 3. *A sequence $x \in L^\infty, L = \{0, 1\}$ is called Kolmogorov-Chaitin random if it is incompressible (no initial segment of x can be compressed more than for a fixed finite number of bits) or in other words:*

$$\exists b > 0 : \tilde{K}(x_{1:n}) \geq n - b \text{ for all } n. \tag{2.5}$$

Equivalence of Kolmogorov-Chaitin and Martin-Löf approaches to randomness

At the first sight the Kolmogorov-Chaitin and Martin-Löf approaches to randomness differ crucially. The first one is about randomness of an individual sequence. There is no reference to other sequences; one is not interested in how typical is this concrete sequence in an ensemble of all possible sequences. We can say that Kolmogorov-Chaitin randomness is determined intrinsically and Martin-Löf randomness is determined externally. Therefore the following result is really surprising:

Theorem 6. (Schnorr [227]) *A sequence is Martin-Löf random if and only if it is Kolmogorov-Chaitin random.*

Coupling between Kolmogorov-Chaitin, Martin-Löf and Mises-Wald-Church randomnesses

Theorem 7. (Invariance of randomness with respect to place selections) *Let $x = (x_j)$ be a Kolmogorov-Chaitin (Martin-Löf) random sequence. Given an algorithmically computable place selection, see (1.49), (1.50), let n_1 be the least n such that $f_n(x_1, ..., x_{n-1}) = 1$, n_2 be the next such n, etc. Then $x = (x_{n_k})$ is also a Kolmogorov-Chaitin (Martin-Löf) random sequence.*

It is also possible to prove that any Kolmogorov-Chaitin (Martin-Löf) random sequence satisfies the principle of statistical stabilization, i.e., relative frequencies tend to limits - probabilities. Finally, we obtain the following important result providing a partial connection with von Mises' notion of collective (with Church's flavor):

Corollary 1. *Any Kolmogorov-Chaitin (Martin-Löf) random sequence is also Mises-Wald-Church random.*

Of course, the inverse is not correct: as was shown by Ville, there exist Mises-Wald-Church random sequences which do not pass the basic statistical tests.

2.4   Randomness: Concluding Remarks

In the community working on foundations of randomness the assertion that *Martin-Löf randomness or equivalently Kolmogorov-Chaitin randomness captures the "true notion of randomness"' conforming to our intuition is sometimes called the Martin-Löf-Chaitin thesis.* (The Martin-Löf-Chaitin thesis, like the Church-Turing thesis for the definition of algorithm, is not a mathematical proposition that can be proved or refuted.) Thus in this community the problem of formalization of the intuitive notion of randomness is considered to be solved. Personally I do not think so.

Of course, the realization of the Kolmogorov-Solomonoff-Chaitin program on randomness as complexity-incompressibility was one of the most important contributions to theory of randomness. Now we have a rigorous mathematical theory of individual random sequences. In the same way the Martin-Löf theory of algorithmic statistical tests provides a rigorous mathematical description of randomness as typicality. This theory establishes the solid theoretical foundation for the widely applicable method of testing of (pseudo)random generators, including the NIST test. An output of a (pseudo)random generator has to pass a block of algorithmically designed tests to be "recognized" as a (pseudo)random sequence. Schnorr's proof that the Kolmogorov-Solomonoff-Chaitin and Martin-Löf approaches match perfectly can be considered as culmination of development of theory of randomness in the 20th century.

Starting with 1960s (and even earlier with the works of Turing) the evolution of mathematical theory of randomness was closely connected to the computer revolution. Development of the art of programming excited people about the idea of algorithmic computability. In the light of this revolution[6]. It was fashionable to formulate the problem of randomness with

---

[6]It was not only computational and technological revolution. It changed drastically the philosophical basis of the modern science. In particular, G. Chaitin remarked [55], p. xiii: "I was fascinated by computers as a child. First of all, because they were a great toy, an infinitely malleable artistic medium of creation. I loved programming! But most of all, because the computer was (and still is!) a wonderful new **philosophical** and mathematical concept. The computer is even more revolu-

the aid of language of computability. However, nowadays when the use of advanced computer programs became the everyday routine and people are not blindfolded by the light of programming anymore, one can ask honestly whether the whole project of the algorithmically based randomness was really so much justified. The dream about creation of computer-like artificial intelligence evaporated. Mind seems not to be driven by computer programs. R. Penrose [216] rightly pointed to the transcendental nature of human mind. In this new context one can try to start a new project of formalization of randomness which is not based on algorithmic computability.

The situation is worse for the approach originated by von Mises (while its importance was recognized already by Laplace) and based on the interpretation of randomness as unpredictability. On one hand, the mathematically rigorous formalization of von Mises notion of collective suffers Ville's objection. On the other hand, in the Kolmogorov-Solomonoff-Chaitin-Martin-Löf framework one can only be sure that a random sequence is unpredictable in the sense of Mises-Church, but not vice versa.

We remark that randomness as unpredictability seems to be the most important for applications. In principle, user is not interested explicitly in complexity of an output of a (pseudo)random generator or whether this output passes a block of algorithmic tests. (We remark that the sequence x_{rand} from section 2.1 composed of the digits of the number π passes the majority of standard tests of randomness. One even had to develop a special test, the so-called π-test, in order to block usage of this sequence.) We are interested only in a possibility to predict the output or, at least, to find some patterns in it. However, as we have seen, the direct formalization of randomness-unpredictability has not yet been created. The von Mises approach (or something totally novel) still waits for its time to come. We remark (private communication[7]) that Kolmogorov died with the hope that

---

tionary as an **idea**, than it is as a practical device which alters the society – and we all know how much it has changed our lives. Why do I say this? Well, the computer changes epistemology, it changes the meaning of "to understand". To me you understand something only if you can program it. (You, not somebody else!) Otherwise you do not really understand it, you only think you understand it." And then he continued: "The computer has provoked a paradigm shift; it suggests a digital philosophy, it suggests a new way of looking at the world, in which everything is discrete and nothing is continuous, in which everything is digital information, 0's and 1's." In physics this paradigm shift led to digital and information physics culminating in Wheeler's "it from bit"." We remark that the process of diffusion of the new digital paradigm from mathematics and computer science to physics was not instantaneous: Wheeler formulated his principle only in 1990 [257].

[7]This is the reply to my question asked during the talk of A. N. Shiryaev at Mathe-

in future a new and unexpected approach to the notion of randomness will be elaborated. And, of course, mathematicians will continue to work on this problem. However, it may happen that future attempts will never lead to a mathematically acceptable notion of randomness.

This was the final point of my second lecture given at IQOQI. In the after-talk discussion various opinion were presented; in particular, Prof. Zeilinger conjectured that such a painful process of elaboration of the mathematical theory of randomness is simply a consequence of the methodological mistreatment of this notion. It might be that *randomness is not mathematical, but a physical notion*. Thus one has to elaborate physical procedures guaranteeing randomness of experimentally produced sequences and not simply try to construct such procedures mathematically.

In some sense Zeilinger's proposal is consonant with von Mises' proposal: *to find a collective, one simply has to go to a casino*. Zeilinger proposes to go to his quantum optics lab at Boltzmanngasse 3.

---

matical Institute of Russian Academy of Science, "Probability and the concept of randomness", 26 November, 2009. When A. N. Kolmogorov became physically disable (but mentally he was still very bright), A. N. Shiryaev spent with him a few years, reading him scientific papers and discussing various problems, in particular, on foundations of probability and randomness.

Chapter 3

Supplementary Notes on Measure-theoretic and Frequency Approaches

3.1 Extension of Probability Measure

This section has two aims: a) to provide a deeper insight to measure-theoretic features of probability; b) to show complexity of the measure-theoretic approach to probability.

We start with the remark that typically it is not easy to "visualize" a σ-algebra of sets $\mathcal{F}$ and one starts with some simply describable system of sets, say $\mathcal{F}_0$, and considers the σ-algebra generated from it with the aid of Boolean operations: $\mathcal{F} = \sigma(\mathcal{F}_0)$. Thus $\mathcal{F}$ is the minimal σ-algebra containing $\mathcal{F}_0$. In the non-discrete case a probability measure P is also explicitly defined only on $\mathcal{F}_0$. And the condition of σ-additivity is explicitly testable only on $\mathcal{F}_0$: equality (1.5) holds for any sequence (A_i) of disjoint sets belonging $\mathcal{F}_0$ such that $A = \cup_{i=1}^{\infty} A_i$ also belongs to $\mathcal{F}_0$. Typically one can assume that $\mathcal{F}_0$ is an algebra of sets.

3.1.1 *Lebesgue measure on the real line*

For example, for $\Omega = [R_1, R_2] \subset \mathbf{R}$, where R_1, R_2 are finite, the *Borel σ-algebra* $\mathcal{B} \equiv \mathcal{B}_{[R_1,R_2]}$ is generated by the intervals of the following forms:

$$I = [a, b], (a, b), [a, b), (a, b], \tag{3.1}$$

where $R_1 \leq a \leq b \leq R_2$. (In fact, the generating system of intervals can be reduced to the intervals of the form $[R_1, b)$.)

Denote the collection of finite unions of disjoint intervals of forms (3.1) by $\mathcal{F}_0$. This is an algebra. Its elements have the form

$$A = \cup_{i=1}^{M} I_i, I_i \cap I_j = \emptyset, i \neq j, \tag{3.2}$$

where (I_i) are the intervals of forms (3.1). Our aim is to define a measure on $\mathcal{B}$ (in fact, on σ-algebra, which is larger) starting with the length of interval. This measure is known as the *Lebesgue measure*. For any interval I, we set

$$\mu(I) = b - a, \tag{3.3}$$

the length of the interval; we can also consider the probabilistic version of μ

$$P(I) = \frac{b - a}{R_2 - R_1}, \tag{3.4}$$

the normalized length of the interval.

First we extend μ on $\mathcal{F}_0$ by setting, for A of form (3.2),

$$\mu(A) = \sum_{i=1}^{m} \mu(I_i). \tag{3.5}$$

It is straightforward to show that μ is σ-additive on $\mathcal{F}_0$, i.e., if $A \in \mathcal{F}_0$ and it can be represented as

$$A = \cup_{k=1}^{\infty} A_k, A_k \in \mathcal{F}_0, A_k \cap A_m = \emptyset, k \neq m,$$

then

$$\mu(A) = \sum_{k=1}^{\infty} \mu(A_k).$$

We remark that here A as well as all A_k can be represented as finite unions of disjoint intervals. Below we present the abstract scheme which can be applied to the Lebesgue measure. In fact, historically this scheme was first elaborated for this concrete measure and then generalized to an arbitrary σ-additive measure.

Since this book is about probability, we shall present the extension scheme directly for probability measures; in particular, see (3.4) for the probability corresponding to the Lebesgue measure on an interval. The same construction can be used for an arbitrary σ-additive measure μ (with minor modifications taking into account that μ is not normalized by 1).

3.1.2   *Outer and inner probabilities, Lebesgue measurability*

Now we turn to the abstract framework. Suppose that $\mathcal{F}_0$ is an algebra and that $P : \mathcal{F}_0 \to [0, 1]$ is σ-additive on it. Can it be extended to a σ-additive probability measure defined on some σ-algebra? The answer is positive. We remark that, since any σ-algebra containing $\mathcal{F}_0$ also contains the σ-algebra $\mathcal{F} = \sigma(\mathcal{F}_0)$ generated by $\mathcal{F}_0$, the aforementioned extension of P automatically covers $\mathcal{F}$. In particular, for the Lebesgue measure, we automatically get its extension to the Borel σ-algebra $\mathcal{B}$.

For any subset A of Ω, we introduce its *outer probability* as

$$P^*(A) = \inf_{A \subseteq \cup B_i, B_i \in \mathcal{F}_0} \sum_i P(B_i), \tag{3.6}$$

where (B_i) is a finite or countable covering of A. This is an abstract scheme corresponding to the process of calculation of the area of a complex figure A by covering it by, e.g., rectangles.

We remark that in general P^* is not additive (e.g., in the case of the Lebesgue measure), i.e., in general, for two disjoint sets A_1 and A_2,

$$P^*(A_1 \cup A_2) \neq P^*(A_1) + P^*(A_2). \tag{3.7}$$

However, it follows directly from the definition that it is *subadditive*, i.e., for any pair of sets A_1, A_2,

$$P^*(A_1 \cup A_2) \leq P^*(A_1) + P^*(A_2). \tag{3.8}$$

Moreover, it is even σ-subadditive, i.e., for any sequence of sets (A_i),

$$P^*(\bigcup_i A_i) \leq \sum_i P^*(A_i). \tag{3.9}$$

Remark 1. In principle, one can proceed quite far with a generalized (non-Kolmogorovian) probability model in which the condition of σ-additivity is modified to the condition of subadditivity. In particular, such "probabilities" are widely used in psychology [106]. However, now our aim is to construct the Kolmogorov probability space starting with the outer probability.

In section 3.2, we use the following property of the outer probability. Let A, B be arbitrary subsets of Ω and let $A \subset B$. Then, cf. (1.9),

$$P^*(A) \leq P^*(B). \tag{3.10}$$

This is a direct consequence of the definition of the outer probability.

For any subset A of Ω, we introduce its *inner probability* as

$$P_*(A) = 1 - P^*(\complement A). \tag{3.11}$$

By using the definitions and the properties of the inf-operation we obtain that

$$P_*(A) \leq P^*(A). \tag{3.12}$$

Since the minimal covering of $A \in \mathcal{F}_0$ by a sequence of subsets belonging $\mathcal{F}_0$ is given by just A, by taking into account (1.11) we obtain:

$$P^*(A) = P(A), \ A \in \mathcal{F}_0. \tag{3.13}$$

By taking into account (1.12) we also obtain that, for $A \in \mathcal{F}_0$,

$$P_*(A) = P(A), \ A \in \mathcal{F}_0. \tag{3.14}$$

Thus both P^* and P_* are extensions of P originally defined on $\mathcal{F}_0$.

Now we introduce the following system of subsets of Ω :

$$\mathcal{F}_L = \{A : P^*(A) = P_*(A)\}. \tag{3.15}$$

As we have seen, $\mathcal{F}_0 \subset \mathcal{F}_L$. We set

$$P_L(A) = P^*(A) = P_*(A), \ \ A \in \mathcal{F}_L. \tag{3.16}$$

Further steps are technically nontrivial; it is possible to show [184] that

- $\mathcal{F}_L$ is a σ-algebra;
- P_L is σ-additive on $\mathcal{F}_L$.

Thus P can be extended from $\mathcal{F}_0$ to the σ-algebra $\mathcal{F}_L$ as a σ-additive probability measure. In particular, we have the following inclusions:

$$\mathcal{F}_0 \subset \mathcal{F} \subset \mathcal{F}_L, \tag{3.17}$$

where as above $\mathcal{F} = \sigma(\mathcal{F}_0)$ is the minimal σ-algebra containing the algebra $\mathcal{F}_0$. Elements of $\mathcal{F}_L$ are called *Lebesgue measurable* sets. Thus probability initially defined on some algebra $\mathcal{F}_0$ of events can be assigned to Lebesgue measurable sets.

The question whether all Lebesgue measurable sets can be really interpreted as events and the extension-measure P_L has the meaning of probability is very complicated. In the measure-theoretic approach it seems natural to work with the Kolmogorov probability space $\mathcal{P}_L = (\Omega, \mathcal{F}_L, P_L)$. It has some features justifying its use in the probability theory.

First of all, P can be extended onto $\mathcal{F}_L$ in the unique way. To formalize this statement, we introduce the following notion. We say that a set A is a set of uniqueness for a σ-additive probability measure on an algebra $\mathcal{F}_0$ if

- There is a σ-additive extension of P defined on A;
- If P_1 and P_2 are two such extensions, then $P_1(A) = P_2(A)$.

It can be shown that the system of sets of uniqueness for σ-additive P coincides with $\mathcal{F}_L$. Hence, by operating with the probability space $\mathcal{P}_L = (\Omega, \mathcal{F}_L, P_L)$ we do not meet the problem of non-uniqueness in assigning probability to an event $A \in \mathcal{F}_L$. Another useful property of $\mathcal{P}_L$ is discussed in section 3.2.

However, von Mises [248], [249] criticized such an approach; he pointed out that, although in formal mathematics one can assign the probability measure P_L to any Lebesgue measurable subset A, from the probabilistic viewpoint the number $P(A)$ might be meaningless. Nowadays von Mises' objection, see also section 3.3, is practically forgotten.

Now consider the problem of the existence of subsets which are not Lebesgue measurable. For example, let $\Omega = [R_1, R_2]$ and let P_L the Lebesgue probability measure. Does $\mathcal{F}_L$ coincide with the σ-algebra of all subsets of Ω? For a physicist, the conventional answer might be unsatisfactory:

NLM *There exist subsets which are not Lebesgue measurable. However, this is the case only if one accepts the validity of the axiom of choice in the set-theory.*

We discuss this axiom and the corresponding proof of the existence of a subset which is not Lebesgue measurable in section 3.4 (the proof is not constructive, it is possible only to prove the existence as a consequence of the axiom of choice)[1]. Thus even by trying to answer such a natural question as **NLM** we confront with the foundations of set-theory which are really nontrivial.

3.2 Complete Probability

The probability measure P_L defined on the σ-algebra of Lebesgue measurable subsets $\mathcal{F}_L$ has one important property. The inequality (3.10) implies that if $A \subset B$ and $P^*(B) = 0$, then also $P^*(A) = 0$. Now let $B \in \mathcal{F}_L, P_L(B) = 0$, and let A be its arbitrary subset. Then A also belongs to $\mathcal{F}_L$, because $P_*(A) \leq P^*(A) \leq P^*(B) = P_L(B) = 0$.

Consider now an arbitrary probability space $\mathcal{P} = (\Omega, \mathcal{F}, P)$. The probability P is called *complete* if $P(B) = 0$ implies that each of its subset A also belongs to $\mathcal{F}$. The latter, of course, implies that $P(A) = 0$.

As we have seen, the probability P_L is always complete (even if the

---

[1]Some mathematicians do not accept such nonconstructive proofs. The corresponding stream is known as constructive mathematics.

original probability was not complete).[2] We remark that any discrete probability measure is complete.

To be complete is very important for a probability measure. This feature is very natural from the probabilistic viewpoint: if the probability of some event B is zero, then the probability of any event composed by a smaller collection of elementary events is also zero.[3] Completeness of a probability measure is also important from the mathematical viewpoint: it plays a crucial role in many theorems about convergence of sequences of random variables.

Initially Kolmogorov missed the issue of completeness but later recognized its importance. We point out that one can easily complete any probability P on a σ-algebra $\mathcal{F}$. It is sufficient to add to $\mathcal{F}$ all subsets of events of zero probability and then consider the σ-algebra generated by this extended system of subsets. The elements of the new σ-algebra, denoted as $\bar{\mathcal{F}}$, have the form

$$C' = C \cup A, \tag{3.18}$$

where $C \in \mathcal{F}$ and A is a subset of an event of zero probability. We set $\bar{P}(C') = P(C)$. Then $\bar{P}$ is a complete probability measure on $\bar{\mathcal{F}}$. The probability space $\bar{\mathcal{P}} = (\Omega, \bar{\mathcal{F}}, \bar{P})$ is called the completion of the probability space $\mathcal{P} = (\Omega, \mathcal{F}, P)$. Finally, we remark that the probability space $\mathcal{P}_L = (\Omega, \mathcal{F}_L, P_L)$ can be interpreted as the completion of the probability space $\mathcal{P} = (\Omega, \mathcal{F}, P)$, where $\mathcal{F} = \sigma(\mathcal{F}_0)$ and P is the restriction of P_L on $\mathcal{F}$. Each Lebesgue measurable set can be represented as (3.18). In particular, for Lebesgue probability measure, each Lebesgue measurable set can be represented as the union of a Borel set and a subset of a Borel set of zero measure. Thus the whole theory about Lebesgue measurability as merely about events of zero probability.

Remark 2. Since in this book we question everything, the assumption of completeness can also be questioned. Heuristically it is not evident that any subset of an event of zero probability should also be recognized as an event. In particular, the wish to proceed with a complete probability measure stimulates the use of the Lebesgue probabilities. And we question this approach in section 3.3.

---

[2] Consider the probability space $([R_1, R_2], \mathcal{B}_{[R_1, R_2]}, P)$, where P is Lebesgue probability measure on the Borel σ-algebra. Here P is not complete. For example, one can find a non-Borel subset of the Cantor set (and the latter has the Lebesgue measure zero).

[3] However, there is some ambiguity in the last statement: in the Kolmogorov theory only the elements of the given σ-algebra $\mathcal{F}$ (incorporated in the definition of the probability space) have the meaning of events.

3.3   Von Mises Views

3.3.1   *Problem of verification*

One of the basic methodological problems of von Mises' approach to probability is the *problem of verifications*. Verification of any of his fundamental principles, the principles of statistical stabilization and randomness, demands infinitely many trials. Hence, such verification is in principle impossible for any concrete experimentally produced random sequence, a collective. We remark that this is just another representation of the problem of verification of σ-additivity in Kolmogorov's approach, see Remark 1 section 1.2.2. However, in some sense von Mises' approach has an important advantage over Kolmogorov's approach. Opposite to Kolmogorov and the majority of scientists who contributed to foundations of probability theory, von Mises considered this theory not as a pure mathematical theory, but as a special theory of natural and social phenomena. He often compared probability theory with hydrodynamics. The latter is not reduced to hydrodynamical equations and their beauty does not shadow the physical content of hydrodynamics. In the same way the beauty of the measure theory shall not shadow the physical or sociological content of the probability theory, as well as theory of randomness which in von Mises' model was irreducibly embedded in the theory of probability. (The debate between von Mises and Hausdorff representing the positions of applied and pure mathematicians with respect to the notion of probability was excellently enlightened in the historical paper of Siegmund-Schultze [234].)

Therefore von Mises was fine with a possibility of *approximate verification*. To verify the probability of some event A, he proposed to perform sufficiently many trials of a random experiment generating A. Of course, here the crucial point is the interpretation of "sufficiently many". However, again von Mises was not upset about this problem. He was sure that it can be properly solved for each experimental situation. For simplicity reasons, we modify the original example of von Mises and consider coin tossing, instead of dice throwing [249]. We consider the problem of the verification of zero probability of the event "in a sequence of coin tossing head (H) will never appear." Following von Mises, one, say Alice, performs, e.g., $n = 1000$ trials. Alice would see that H will appear a few times. Then Alice performs $n = 10000$ trials and if she is still unsatisfied (and has time to continue experimenting), then she can take $n = 100000$.

In spite of the belief that approximate verification is the best possible scientific methodology for probability theory[4], von Mises also had a problem with general applicability of this methodology. He understood that some events cannot be verified, even approximately. Consider the problem of verification for the probability the event A_{fnt} that in the continuing tossing of a coin H appears a finite number of times only. Alice can make $n = 1000$, or $n = 10000$, or even $n = 1000000$ trials. She can never be even approximately sure that the event A_{fnt} occurred or not. Such example led von Mises to understand that there exist events for which probabilities are not verifiable, even approximately.

Now we compare von Mises' and Kolmogorov's approaches. Consider some probability space $\mathcal{P} = (\Omega, \mathcal{F}, P)$. Here only subsets A of Ω belonging to the specially selected system $\mathcal{F}$ are recognized as events and, hence, probabilities can be assigned only to them. However, the system $\mathcal{F}$ has to be a σ-algebra, so it has to be sufficiently extended. From von Mises' viewpoint, this mathematical requirement does not match real probabilistic situation.

Consider, for example, the Kolmogorov probability space such that the sample space $\Omega = [0,1]$ and $\mathcal{F} = \mathcal{B}_{[0,1]}$, the Borel σ-algebra, and P is the Lebesgue probability measure, see section 3.1.1. We remark that $\mathcal{B}_{[0,1]}$ is the minimal σ-algebra containing such natural evens as intervals; Kolmogorov could not proceed with a smaller family of events. However, for von Mises, even some events represented by the Borel subsets are not (even approximately) verifiable.

For example, the event A_{fnt} can be represented as a Borel subset of $[0,1]$. Consider the representation of real numbers with respect to base two, i.e., as sequences of digits from $\{0,1\}$. Then the event A_{fnt} can be represented as the subset of $[0,1]$ such that each of its element $\omega = (\alpha_1...\alpha_n....), \alpha_j = 0,1$, contains only a finite number of zeros (representing the appearance of H). This is a Borel set and its Lebesgue measure equals zero (the reader can try to prove this). Hence, from Kolmogorov's viewpoint A_{fnt} is just one of events of zero probability. However, von Mises said that assigning probability to A_{fnt} is meaningless.

In mathematical terms von Mises disagreed with the use of σ-algebras for the mathematical representation of events. Instead of the Lebesgue approach to measurability which led to σ-algebras, section 3.1.2, he explored the Jordan approach to measurability which led to algebras of events.

---

[4]Cf. with Popper's methodology of falsification of a scientific theory; see, especially, [100].

3.3.2 *Jordan measurability*

Let μ be a measure (in general, finite additive) on some set-algebra $\mathcal{F}_0$. For example, the Lebesgue measure on the algebra $\mathcal{F}_0$ generated by intervals (each element of it can be represented as union of finitely many disjoint intervals). Let A be an arbitrary subset of Ω. It is said to be *Jordan measurable*, if for every $\epsilon > 0$, there are $A_-, A_+ \in \mathcal{F}_0$, such that

$$A_- \subset A \subset A_+, \; \mu(A_+ \setminus A_-) < \epsilon. \tag{3.19}$$

It can be shown that the collection of all Jordan measurable subsets $\mathcal{F}^*$ is an algebra and it contains the original algebra $\mathcal{F}_0$. The procedure of extension of μ from $\mathcal{F}_0$ to $\mathcal{F}^*$ is similar to the procedure based on the notion of Lebesgue measurability. The main difference is that in exploring the latter we were able to preserve σ-additivity, but in the coming construction the output measure can be finite additive even if the input measure on $\mathcal{F}_0$ was σ-additive, as in the case of the Lebesgue measure.

Let A be an arbitrary subset of Ω. We set

$$\mu_+(A) = \inf_{B \in \mathcal{F}_0, A \subset B} \mu(B), \tag{3.20}$$

and

$$\mu_-(A) = \sup_{B \in \mathcal{F}_0, B \subset A} \mu(B). \tag{3.21}$$

Then it can be shown that

$$\mathcal{F}^* = \{A : \mu_+(A) = \mu_-(A)\}. \tag{3.22}$$

For $A \in \mathcal{F}^*$, the Jordan measure is defined as

$$\mu(A) = \mu_+(A) = \mu_-(A).$$

This is a measure on $\mathcal{F}^*$. If μ is σ-additive on $\mathcal{F}_0$, then its Jordan extension is σ-additive as well, but $\mathcal{F}^*$ need not be a σ-algebra.

It is interesting to establish relation between the Jordan algebra $\mathcal{F}^*$ and the Lebesgue σ-algebra $\mathcal{F}_l$. We have

$$\mathcal{F}^* \subset \mathcal{F}_L \tag{3.23}$$

and for $A \in \mathcal{F}^*$, the Jordan and Lebesgue measures coincide. The inclusion (3.23) is proper, i.e., there exist sets which are Lebesgue measurable, but not Jordan measurable.

We remark that the construction of the Lebesgue measure on the real line can be easily generalized to $\mathbf{R}^n$, by considering the direct products of n intervals as the starting set system $\mathcal{F}_0$. The Borel σ-algebra is generated by $\mathcal{F}_0$ and the σ-algebra $\mathcal{F}_L$ of Lebesgue measurable sets is constructed as usual. Then *a set A is Jordan measurable if and only if the Lebesgue measure of its boundary equals zero.* For example, the subset of rational numbers of a segment $[R_1, R_2]$ is not Jordan measurable, because the Lebesgue measure of its boundary equals to 1. We remark that this set belongs even to the Borel σ-algebra, i.e., $\mathcal{B} \not\subset \mathcal{F}^*$. The system of Jordan measurable sets is intrinsically an algebra – there is no σ-algebra inside it.

For von Mises, this characterization of Jordan measurable sets was one of the motivations to consider them and only them as mathematical representatives of events. Von Mises thought that if a measure of the boundary of some set A is not zero, then we cannot ignore the probabilistic weight of its boundary points. But, for boundary points, it is difficult to verify whether they contribute to the "event" A or its negation CA.

3.4   Role of the Axiom of Choice in the Measure Theory

We shall use the notion of a *binary relation*. Let M be a set and let R be a subset of the Cartesian product $M \times M$, i.e., some collection of ordered pairs (a, b) of elements of M. Then we say that a is related to b by the binary relation $R : aRb$. A relation between elements of M is called a *relation on M* if, for every $a \in M$, there is at least one ordered pair (a, b), i.e., $(a, b) \in R$.

Axiom of choice. *Given any set M, there exists a "choice function" f such that $f(A)$ is an element of A for any nonempty subset A of M.*

The validity of this axiom is typically assumed automatically in any set-theoretic framework. However, from the viewpoint of foundations of the set theory the use of this axiom generates numerous controversies.

Example of non-measurable set

Let Ω be a circle of the radius 1. As probability we select the Lebesgue measure μ on Ω. (To be more precise, we have to speak about the Haar measure μ on the circle which is considered as the group of rotations. This measure is rotationally invariant: for any measurable subset A, rotation by any angle ϕ does not change the measure of A.)

Let α be an irrational number. Consider on the circle the equivalence relation: two points are equivalent if one can be obtained from the other by rotating the circle through an angle $\phi_n = \pi \alpha n$ $(n = 0, \pm 1, \pm 2, ...)$. As any equivalence relation, this relation splits Ω into disjoint classes. Each class contains countably many points. The crucial point of the existence proof is the application of the Axiom of choice. Let A_0 be any set which is obtained by selecting one point from each equivalence class. It can be shown that A_0 is not Lebesgue measurable, $A_0 \notin \mathcal{F}_L$.

To prove non-measurability of A_0, we consider sets A_n obtained by rotating A_0 by the angle ϕ_n. Then

$$\Omega = \cup_{n=-\infty}^{\infty} A_n, \; A_n \cap A_m = \emptyset, n \neq m.$$

If A_0 were measurable, the congruent sets A_n would also be measurable. Then the σ-additivity of the measure μ on the circle would imply that

$$1 = \mu(\Omega) = \mu(\cup_{n=-\infty}^{\infty} A_n) = \sum_{n=-\infty}^{\infty} \mu(A_n). \qquad (3.24)$$

However, congruent sets must have the same measure, i.e., if A_0 were measurable, then it should be that

$$\mu(A_n) = \mu(A_0),$$

which contradicts (3.24).

3.5 Possible Generalizations of Probability Theory

Kolmogorov himself did not consider his axiomatics as something final and unchangeable. He discussed various possible modifications of the axiomatics. In particular, for him the requirement that the collection of all possible events has to form a σ-algebra or even an algebra of sets was questionable. He proposed some models of "generalized probability" where the collection of events need not have the standard structure of (σ-)algebra.

However, later the modern axiomatics of probability was so to say "crystallized" and any departure from the Kolmogorov axiomatics was considered as a kind of pathology.

Nevertheless, the concrete applications continuously generate novel models in which probabilities have some unusual features. For example, in psychology and game theory non-additive "probabilities" were invented and actively used, see, e.g., [163] for references. There are also "negative probabilities" which were explored in the works of Dirac [77], Feynman [89], Aspect [19] and my own works [156]; see also Mückenheim [205] for a detailed review on negative probabilities in quantum physics and recent paper of Oas et al. [209]. Dirac even used "complex probabilities" [77].

3.5.1 *Negative probabilities*

"Negative probabilities" appear with strange regularity in a majority of problems in quantum theory starting with Dirac relativistic quantization of the electromagnetic field. Feynman was sure that all quantum processes can be written with the aid of negative transition probabilities which disappear in the final answers corresponding to the results of measurements. Of course, the interpretation of signed measures taking, in particular, negative values, as probabilities (in particular, the use of signed density functions) is counterintuitive. Nevertheless, operations with such generalized probabilities are less counterintuitive than operations in the formal quantum framework. One can say (as Feynman often did) that the only difference between classical and quantum physics is that in the latter one has to use generalization of the Kolmogorov measure-theoretic model with signed measures, instead of positive measures.

The mathematical formalism of theory of negative probabilities was presented in my book [133] (and its completed edition [156]). Mathematically this is a trivial generalization of the Kolmogorov notion of probability space.

Let again Ω be a set and let $\mathcal{F}$ be a σ-algebra of its subsets. A *signed probability measure* P is a map from $\mathcal{F}$ to the real line $\mathbf{R}$, normalized $P(\Omega) = 1$ and σ-additive. The *signed probability space* [156] is a triple $\mathcal{P} = (\Omega, \mathcal{F}, P)$. Points of Ω are said to be *elementary events*, elements of $\mathcal{F}$ are *random events*, P is *signed probability*.

If such a "probability" takes negative values, then automatically it also has to take values exceeding 1. It is easy to show. Take an event A such that $P(A) < 0$. Take its complement $B = \Omega \setminus A$. Then additivity and normalization by 1 imply that $P(\Omega) = P(A \cup B) = P(A) + P(B) = 1$, i.e., $P(B) > 1$.

On the basis of the signed probability space it is possible to advance very far by generalizing the basic notions and constructions of the standard probability theory: conditional probabilities are defined by the Bayes formula, independence is defined through factorization of probability, the definition of a random variable is the same as in the standard approach, the average (mean value) is given by the integral.[5]

The complex probability space is defined in the same way; here "probabilities" are normalized σ-additive functions taking values in the field of

---

[5] We remark that each signed measure μ can be canonically represented as the difference of two positive measures (*the Jordan decomposition of a signed measure*): $\mu = \mu_+ - \mu_-$. Thus one can define the integral with respect to a signed measure as the difference of integrals with respect to these two positive measures.

complex numbers **C**.

The prejudice against such generalized probabilities is the result of a few centuries of use of the Laplacian definition of probability as the proportion $P(A) = m/n$, where m is the number of cases favorable to the event A and n is the total number of possible cases. Take, for example, a coin, then here $n = 2$ and, for the event H that the head will appear in a particular trial, $m = 1$, $P(H) = 1/2$. For a dice, we shall get probabilities $P(A_j) = 1/6$, where the index $j = 1, 2, ..., 6$ labels the corresponding sides of the dice. It is clear that the Laplacian probability cannot be negative or exceed 1. However, the domain of application of the Laplacian probability is very restricted. For example, it can be used only for symmetric coins and dices. (Nevertheless, since the Laplacian probability is the starting point of all traditional courses in probability theory, the notion of probability is often associated precisely with its Laplacian version. In particular, this image of probability was strongly criticized by Richard von Mises [248, 249].) If one assign weights to events by using not only the Laplacian rule (which is applicable only in very special cases), then, in principle, one may try to use any real number as the weight of an event. There is an order structure on the real line. Thus it is possible to compare probabilities and to say that one event is more probable than the other. Hence, the (b)-part of the Kolmogorov interpretation of probability, see section 1.8, is also applicable to signed probabilities.

For signed probabilities, the possibility to use the frequency interpretation of probability, the (a)-part of the Kolmogorov interpretation (see section 1.8), is a more delicate question. Surprisingly there were obtained analogues of the law of large numbers and the central limit theorem.[6] However, it is impossible to obtain such a strong type of conver-

---

[6] It is convenient to proceed with complex "probabilities", i.e., to consider the signed probabilities a particular case of complex probabilities, see [132] for the most general case: the limit theorems were derived for "probabilities" taking values in complex Banach (super)algebras. This possibility to generalize the basic limit theorems of the standard probability theory to real and complex valued probability measures (as well as to more general measures) is also a supporting argument to use, e.g., negative probabilities. Sometimes mathematics can lead us to areas where our intuition does not work. To illustrate the common prejudice against the use of signed and complex probabilities, I present the following story from time of my PhD-studies. A. N. Kolmogorov recommended my paper (joint with my PhD-supervisor O.G. Smolyanov) about the central limit theorem for complex-valued probabilities to Doklady USSR [236]. He felt very positively about such a generalization of his theory of probability. However, the editors of Doklady sent this paper to two additional reviewers who both wrote negative reports motivating their opinions by total meaninglessness of attempts to generalize Kolmogorov's axiomatics. Finally, the absurdness of the situation became clear even to the editors of Doklady and

gence as we have for the standard positive probabilities. For example, in the generalized law of large numbers the arithmetic mean need not converge to the probabilistic mean m almost everywhere. For identically distributed and independent random variables $\xi_1, ..., \xi_N, ...$, with the mean value $m = E\xi_j = \int_\Omega \xi_j(\omega)dP(\omega)$, it can happen that

$$P\left(\omega \in \Omega : \frac{\xi_1(\omega) + ... + \xi_N(\omega)}{N} \to m, N \to \infty\right) \neq 1. \qquad (3.25)$$

Nevertheless, a weaker convergence is still possible. Set

$$\eta_N = \frac{\xi_1(\omega) + ... + \xi_N(\omega)}{N}.$$

Then it is possible to show that, for a sufficiently extended class of functions, $f : \mathbf{R} \to \mathbf{R}$, the following form of the weak convergence takes place:

$$Ef(\eta_N) = \int_\Omega f(\eta_N(\omega))dP(\omega) \to f(m), N \to \infty. \qquad (3.26)$$

We remark that, for the standard positive probability, the limit relation (3.26) holds for all bounded continuous functions (this is a consequence of the strong law of large numbers). For generalized, signed or complex, probabilities (or even probabilities with values in Banach algebras) functions for which (3.26) holds must have some degree of smoothness [132].

3.5.2 *On generalizations of the frequency theory of probability*

Another way to obtain the frequency interpretation of *negative probabilities* is to use the number-theoretic approach based on the so-called p-adic numbers. The starting point is that frequencies $\nu_n = n/N$ are always *rational numbers*. We recall that the set of rational numbers $\mathbf{Q}$ is a dense subset in the set of real numbers $\mathbf{R}$, i.e., the real numbers can be obtained as limits of rational numbers. Von Mises implicitly explored this number-theoretic fact in the formulation of the principle of the statistical stabilization of relative frequencies. For example, the probability of an event is a real number (in $[0, 1]$), because he considered one special convergence on $\mathbf{Q}$, namely, with respect to the metric induced from $\mathbf{R} : \rho_{\mathbf{R}}(x, y) = |x - y|$. The additivity of the frequency probability is the result of the continuity of the operation of addition with respect to this metric. It can be shown [126] that to get the Bayes formula, one has to use the continuity of the operation of division

the paper was published.

in $\mathbf{R} \setminus \{0\}$. Thus the von Mises theory of frequency probability is based on the dense embedding

$$\mathbf{Q} \subset \mathbf{R}. \tag{3.27}$$

It is important that $\mathbf{R}$ is a *topological field*, i.e., all algebraic operations on it are continuous.

This analysis of the number-theoretic-topological structure of the von Mises model leads us to the following natural question:

Is it possible to find dense embedding of the set of rational numbers $\mathbf{Q}$ (representing all relative frequencies) into another topological number field K, $\mathbf{Q} \subset K$?

If the answer were positive, then one would be able to extend the domain of application of the principle of the statistical stabilization of von Mises: to consider the problem of the existence of the limits of relative frequencies in the topology of the number field K. Since the standard properties of probability were based on the consistency between the algebraic and topological structures, one can expect that, for a topological field, the corresponding generalization of the (frequency) probability theory will be very similar to the standard one based on the embedding (3.27).

3.5.3 *p-adic probability*

It is interesting that there are not so many ways to construct dense embeddings of $\mathbf{Q} \subset K$; in other words: to complete the field of rational numbers with respect to some metric and to obtain a topological field. By the famous theorem of number theory, *the Ostrowsky theorem*, all "natural embeddings" are reduced to the fields of the so called *p-adic numbers* $\mathbf{Q}_p$ and the field of real numbers $\mathbf{R}$. where $p = 2, 3, ..., 1997, 1999, ...$ are prime numbers. The elements of these fields have the form:

$$a = \sum_{j=k}^{\infty} a_j p^j, a_j = 0, 1, .., p-1,$$

where k is an integer; in particular, for $p = 2$ this is a binary expansion. We recall that each real number can be represented as

$$a = \sum_{-\infty}^{k} a_j p^j, a_j = 0, 1, .., p-1.$$

Thus a real number can contain infinitely many terms with the negative powers of p and a p-adic number can contain infinitely many terms with the positive powers of p.

In principle, for some sequences $x = (x_1, ..., x_n, ...)$ of the results of trials (where $x_j \in L$ and L is the corresponding label set), the limits of relative frequencies may exist not in $\mathbf{R}$, but in one of $\mathbf{Q}_p$. Such a "collective" generates probabilities belonging not to $\mathbf{R}$, but to $\mathbf{Q}_p$. One has not to overestimate exoticism of such "probabilities". It is important to understand that the real numbers (opposite to the rational numbers) are just symbols (representing limits of sequence of rational numbers); the p-adic numbers are similar symbols.

Foundations of the p-adic probability theory can be found in [127], [126], [133], [156]. At the beginning this theory was developed in the framework of frequency probability theory. As in the case of ordinary probability theory, the basic properties of the frequency probability were used as axioms of measure-theoretic theory of probability (cf. with Kolmogorov's use of von Mises theorems as axioms [177]). One of the interesting mathematical facts about such generalized probabilities is that the range of the values of p-adic probabilities coincides with the field of p-adic numbers $\mathbf{Q}_p$, i.e., any p-adic number can be in principle obtained as the limit of relative frequencies, $a = \lim_{N \to \infty} n/N$ (cf. with the standard frequency probability theory of von Mise which implies that the range of values of probability coincides with the segment $[0, 1]$ of the real line). In particular, any rational number (including negative numbers) can be obtained in this way.

Therefore in the p-adic framework "negative probabilities" appear in the natural way. Typically convergence of a sequence of rational numbers (in particular, relative frequencies) in $\mathbf{Q}_p$ implies that this sequence does not converge in $\mathbf{R}$ and vice versa, see [127], [126] for examples. In this framework the appearance of the p-adic and, in particular, negative probabilities, can be interpreted as a sign of irregular (from the viewpoint of the standard probability theory) statistical behavior, as a violation of the basic principle of the standard probability theory: the principle of the statistical stabilization of relative frequencies (with respect to the real metric on $\mathbf{Q}$). In the measure-theoretic approach this situation can be considered as a violation of the law of large numbers.

In short: negative probability means that, for some event A, in the standard decimal (or binary) expansion of the relative frequencies $\nu_N(A) = n(A)/N$ the digits would not stabilize when $N \to \infty$. We remark that, in principle, there is nothing mystical in such behavior of data, see again for examples [127].

Finally, we remark that a p-adic analogue of the Martin-Löf approach to randomness was developed in [133], [156].

3.6   Quantum Theory: No Statistical Stabilization for Hidden Variables?

As will be discussed in Chapter 6, studies on foundations of quantum theory generated a number of deep philosophic problems. One of the basic questions is whether it is possible to use realist models of quantum phenomena. The meaning of the terms realism and reality in the quantum framework will be discussed in very detail in section 6.1.1. Now we just say that realism means that the results of quantum measurements can be treated as the objective properties of a quantum system; for example, electron's spin can be assigned to the electron even in the absence of devices measuring it. This problem is also known as the problem of hidden parameters, i.e., existence of parameters (may be hidden) determining the values of quantum observables before measurements. Nowadays it is commonly accepted that either realism is incompatible with the quantum model, i.e., there are no hidden variables, or such variables are nonlocal. This is a consequence of violation of Bell's inequality (Chapter 8). The statement about Bell's inequality is known as Bell's no-go theorem. As any no-go theorem, it is based on some conditions which can be analyzed and criticized (see [161], [113]). In this section we want to use generalized frequency probability models as a critical argument against Bell's reasoning. (In principle it may be that this section should appear in Chapter 8. However, we want to appeal to the fresh discussion on possible generalization of the frequency definition of probability. In the further chapters it may be shadowed by the quantum formalism and extended discussions on quantum foundations.)

Feynman and Aspect pointed out that in quantum theory hidden variables have to be described by signed probability distributions. In light of the previous considerations this statement can be interpreted in a very natural manner: relative frequencies for hidden variables, $\nu_N(\lambda) = n_N(\lambda)/N$, do not satisfy the principle of the statistical stabilization.

This way the realism can be saved: the local hidden variables are not forbidden, but the principles of the standard probability theory are not applicable to them. In particular, one cannot proceed as Bell did – by operating with the standard (Kolmogorov) probability distribution of hidden parameters. (In works on Bell's inequality it is typically denoted as $\rho(\lambda)d\lambda$, where $\rho(\lambda)$ is the density of hidden variables.)

However, it is not clear whether such a viewpoint leads to reestablishing of realism at the subquantum level. My personal viewpoint is that realism by itself is nothing else but the acceptance of the global validity of the principle of the statistical stabilization (the law of large numbers), i.e., *realism is the notion of statistical nature*. Even statements about validity with probability one or zero are statistical, since they presume repeatability of phenomena under study.

It seems that all our senses were formed to recognize only statistically repeatable phenomena (and what is important: statistically repeatable with respect to the real metric). As a consequence all our measurement devices are designed to observe only such phenomena.[7]

What is still unclear to me is how the continuous space-time structure (modeled with the aid of real numbers) is transferred into the statistical stabilization of frequencies of observed events. It is clear that all our measurement devices explore this space-time structure. Are they just machines which transfer features of the continuous space-time into relative frequencies stabilizing with respect to the real metric and leading to the standard models of probability? Or is it another way around? Do we get the standard model of space-time just because our senses are based on the principle of the statistical stabilization with respect to the real metric? In fact, the convergence with respect to the real metric is based on the agreement that for large N its inverse $1/N$ is a small quantity, if N is very large then $1/N$ is negligibly small and can be neglected. Is this the essence of modern probability theory and even the modern science in general?

---

[7]To be honest, I was not able to find experimental statistical data violating the standard principle of the statistical stabilization. Although in the book [127] the reader can find various examples of stochastic evolution of biological systems in which statistical data violate the principle of the statistical stabilization, these data relates to, so to say, hidden features of such biological evolutions.

Chapter 4

Introduction to Quantum Formalism

In this chapter we present a very brief introduction to the quantum formalism. On one hand, the chapter can serve as the first step towards understanding quantum theory. We present sufficient material to understand essentials of quantum theory and especially its probability structure. On the other hand, the chapter may be of some interest for physicists who know quantum theory very well, since this is quite an unusual introduction to the quantum formalism. To escape mathematical difficulties, we restrict our considerations to the *case of finite dimensional state spaces*. However, the basic definitions and constructions are applicable even for infinite-dimensional state spaces. Interpretational issues will be considered in Chapter 6.

4.1 Quantum States

We start with the definition of pure and mixed quantum states. The state space of a quantum system is based on complex Hilbert space. Denote it by H. This is a complex linear space endowed with a scalar product (a positive-definite non-degenerate Hermitian form) which is complete with respect to the norm corresponding to the scalar product.[1] Denote the scalar product by $\langle \cdot | \cdot \rangle$. In mathematical texts it is typically denoted as $\langle \cdot, \cdot \rangle$. Our notation is close to Dirac's symbolism which is widely used in physics, see section 4.9.

Hilbert space and linear operators. For reader's convenience, we recall that the scalar product is a function from the Cartesian product

---

[1] In the finite-dimensional case the norm and, hence, completeness are of no use. Thus those who have no idea about functional analysis (but know essentials of linear algebra) can treat H simply as a finite-dimensional complex linear space with the scalar product.

$H \times H$ to the field of complex numbers $\mathbb{C}$, $\psi_1, \psi_2 \to \langle \psi_1 | \psi_2 \rangle$, having the following properties:

(1) Positive-definiteness: $\langle \psi | \psi \rangle \geq 0$ with $\langle \psi, \psi \rangle = 0$ if and only if $\psi = 0$.
(2) Conjugate symmetry: $\langle \psi_1 | \psi_2 \rangle = \overline{\langle \psi_2 | \psi_1 \rangle}$.
(3) Linearity with respect to the second argument[2]: $\langle \phi | k_1 \psi_1 + k_2 \psi_2 \rangle = k_1 \langle \phi | \psi_1 \rangle + k_2 \langle \phi | \psi_2 \rangle$, where k_1, k_2 are complex numbers.

It generates the norm on H : $\|\psi\| = \sqrt{\langle \psi | \psi \rangle}$.

A reader who does not feel comfortable in the abstract framework of functional analysis can simply proceed with the Hilbert space $H = \mathbf{C}^n$, where $\mathbf{C}$ is the set of complex numbers, and the scalar product

$$\langle u | v \rangle = \sum_i u_i \bar{v}_i, u = (u_1, ..., u_n), v = (v_1, ..., v_n). \qquad (4.1)$$

In this case the above properties of a scalar product can be easily derived from (4.1). Instead of linear operators, one can consider matrices.

We also recall a few basic notions of theory of linear operators in complex Hilbert space. For a linear operator A, the symbol A^* denotes its *adjoint operator* which is defined by the equality

$$\langle A\psi_1 | \psi_2 \rangle = \langle \psi_1 | A^* \psi_2 \rangle. \qquad (4.2)$$

By fixing in H some orthonormal basis (e_i) and by denoting the matrix elements of the operators A and A^* as a_{ij} and a_{ij}^* we rewrite the definition (4.2) in terms of the matrix elements:

$$a_{ij}^* = \bar{a}_{ji}.$$

A linear operator A is called *Hermitian* if it coincides with its adjoint operator:

$$A = A^\star.$$

If an orthonormal basis in H is fixed, (e_i), and A is represented by its matrix, $A = (a_{ij})$, where $a_{ij} = \langle Ae_i | e_j \rangle$, then it is Hermitian if and only if

$$\bar{a}_{ij} = a_{ji}.$$

We remark that, for a Hermitian operator, all its eigenvalues are real. In fact, this was one of the main reasons to represent quantum observables by Hermitian operators. In quantum formalism the spectrum of a linear operator (the set of eigenvalues while we are in the finite dimensional case)

---

[2]In mathematical texts one typically considers linearity with respect to the first argument. Thus a mathematician has to pay attention to this difference.

coincides with the set of possibly observable values. This is one of the postulates of QM (see, e.g., [161] for the axiomatic formulation of QM). Physicists were sure that results of observations have to be represented by *real numbers* as it was in classical physics, see [127] for critical analysis of this assumption. We also recall that eigenvectors of Hermitian operators corresponding to different eigenvalues are orthogonal. This property of Hermitian operators plays some role in justification of the projection postulate of QM, see again section 4.2.1.

A linear operator A is *positive-semidefinite* if, for any $\phi \in H$,

$$\langle A\phi|\phi \rangle \geq 0.$$

This is equivalent to positive-semidefiniteness of its matrix.

For a linear operator A, its trace is defined as the sum of diagonal elements of its matrix in any orthonormal basis:

$$\mathrm{Tr}A = \sum_i a_{ii} = \sum_i \langle Ae_i|e_i \rangle,$$

i.e., this quantity does not depend on a basis.

Let L be a subspace of H. The orthogonal projector $P : H \to L$ onto this subspace is a Hermitian, idempotent (i.e., coinciding with its square) and positive-semidefinite operator:

(1) $P^* = P$;
(2) $P^2 = P$;
(3) $P \geq 0$.

Here 3) is a consequence of 1) and 2). Moreover, an arbitrary linear operator satisfying 1) and 2) is an orthogonal projector - onto the subspace PH.

Pure states. Pure quantum states are represented by normalized vectors, $\psi \in H : \|\psi\| = 1$. Two colinear vectors,

$$\psi' = \lambda\psi, \lambda \in \mathbf{C}, |\lambda| = 1, \tag{4.3}$$

represent the same pure state. Thus, rigorously speaking, a pure state is an equivalence class of vectors having the unit norm: $\psi' \sim \psi$ for vectors coupled by (4.3). The unit sphere of H is split into disjoint classes - pure states. However, in concrete calculations one typically uses just concrete representatives of equivalent classes, i.e., one works with normalized vectors.

Each pure state can also be represented as the projection operator P_ψ which projects H onto one dimensional subspace based on ψ. For a vector $\phi \in H$,

$$P_\psi\phi = \langle \phi|\psi \rangle \, \psi. \tag{4.4}$$

The trace of the one dimensional projector P_ψ equals 1:

$$\text{Tr } P_\psi = \langle \psi | \psi \rangle = 1.$$

We summarize the properties of the operator P_ψ representing the pure state ψ. It is

(1) Hermitian,
(2) positive-semidefinite,
(3) trace one,
(4) idempotent.

Moreover, any operator satisfying 1)-4) represents a pure state. Properties 1) and 4) characterize orthogonal projectors, property 2) is their consequence. Property 3) implies that the projector is one dimensional.

Mixed states. The next step in the development of QM was the extension of the class of quantum states, from pure states represented by one dimensional projectors to states represented by linear operators having the properties 1)-3).

Such operators are called *density operators.* (This nontrivial step of extension of the class of quantum states was based on the efforts of Landau and von Neumann.) The symbol $D(H)$ denotes the space of density operators in the complex Hilbert space H.

One typically distinguish pure states, as represented by one dimensional projectors, and mixed states, the density operators which cannot be represented by one dimensional projectors. The terminology "mixed" has the following origin: any density operator can be represented as a "mixture" of pure states (ψ_i) :

$$\rho = \sum_i p_i P_{\psi_i}, \; p_i \in [0,1], \sum_i p_i = 1. \tag{4.5}$$

The state is pure if and only if such a mixture is trivial: all p_i, besides one, equal zero. However, by operating with the terminology "mixed state" one has to take into account that the representation in the form (4.5) is not unique. The same mixed state can be presented as mixtures of different collections of pure states.

Any operator ρ satisfying 1)-3) is diagonalizable (even in infinite-dimensional Hilbert space), i.e., in some orthonormal basis it is represented as a diagonal matrix, $\rho_{ij} = \delta_{ij} p_j$, where $p_j \in [0,1], \sum_j p_j = 1$. Thus it can be represented in the form (4.5) with mutually orthogonal one dimensional projectors. The property 4) can be used to check whether a state is pure or not. We point out that pure states are merely mathematical abstractions; in real experimental situations it is possible to prepare only mixed states; one defines the degree of *purity* as $\mathrm{Tr}\rho^2$. Experimenters are satisfied by getting this quantity near one.

Hilbert space of square integrable functions. Although we generally proceed with finite-dimensional Hilbert spaces, it is useful to mention the most important example of infinite-dimensional Hilbert space used in QM. Consider the space of complex valued functions, $\psi : \mathbb{R}^m \to \mathbb{C}$, which are square integrable with respect to the Lebesgue measure[3] on $\mathbb{R}^m$:

$$\|\psi\|^2 = \int_{\mathbb{R}^m} |\psi(x)|^2 dx < \infty. \tag{4.6}$$

It is denoted by the symbol $L^2(\mathbb{R}^m)$. Here the scalar product is given by

$$\langle \psi_1 | \psi_2 \rangle = \int_{\mathbb{R}^m} \bar{\psi}_1(x)\psi_2(x)dx.$$

A delicate point is that, for some measurable functions, $\psi : \mathbb{R}^m \to \mathbb{C}$, which are not identically zero, the integral

$$\int_{\mathbb{R}^m} |\psi(x)|^2 dx = 0. \tag{4.7}$$

We remark that the latter equality implies that $\psi(x) = 0$ a.e. (almost everywhere). Thus the quantity defined by (4.6) is, in fact, not norm: $\|\psi\| = 0$ does not imply that $\psi = 0$. To define a proper Hilbert space, one has to consider as its elements not simply functions, but classes of equivalent functions, where the equivalence relation is defined as $\psi \sim \phi$ if and only if $\psi(x) = \phi(x)$ a.e. In particular, all functions satisfying (4.7) are equivalent to the zero-function.

In the same way one can proceed with L^2-space with respect to an arbitrary σ-additive measure μ defined on some configuration space X which is endowed with some σ-algebra $\mathcal{F}$, the domain of definition of μ. To escape measure-theoretic problems, it is assumed that the measure μ is complete, see section 3.2 (we discussed completeness of probabilistic measures, but the case of an arbitrary measure is handled in the same way).

Here, $\psi : X \to \mathbf{C}$, and

$$\|\psi\|^2 = \int_X |\psi(x)|^2 \mu(dx) < \infty.$$

---

[3]To be completely precise, we have to point that here measurability of ψ is considered with respect to the Lebesgue σ-algebra $\mathcal{F}_L$, see section 3.1.2. But the reader may ignore this technicality.

The scalar product is given by

$$\langle \psi_1 | \psi_2 \rangle = \int_X \bar{\psi}_1(x) \psi_2(x) \mu(dx).$$

We postpone consideration of interpretational questions, including the quantum state interpretations, to Chapter 6. In this chapter we proceed operationally: a quantum state is a mathematical quantity used to determine probabilities for results of measurements, see the Born rule (4.18).

4.2   First Steps Towards Quantum Measurement Theory

Formalization of quantum measurement theory was started by Dirac [76] and von Neumann [250] (with the important contribution of Lüders [194]). The final mathematical formulation is based on the operator representation of physical apparatuses by quantum instruments, see Davis and Lewis [66], Holevo [117], [118], Ozawa [211], [212], Busch, Grabowski, and Lahti [48].

The notion of quantum instrument plays an important role in QM. In this section we present generalization of this notion for an arbitrary state space. Later we shall use it only in QM. However, coming general scheme is useful from the foundational viewpoint. It enlightens connection with formalization of measurement processes in other domains of science, e.g., in psychology. Consider systems of any origin (physical, biological, social, financial). Suppose that the states of such systems can be represented by points of some set X. These are statistical states, i.e., knowing the state of a system we can determine the values of observables only with some probabilities. Suppose that the range of values of an observable a is discrete $O = \{\alpha_1, ..., \alpha_n\}$. Then, for each value α_i of the observable a, there is defined a map

$$p(\alpha_i | x) = f_{a,\alpha_i}(x) \qquad (4.8)$$

giving the probability of the result $a = \alpha_i$ for systems in the state $x \in X$. Here $f_{a,\alpha_i} : X \to [0, 1]$. Then it is natural to assume that the measurement modifies the state x, i.e., there is defined another map

$$x_i = g_{a,\alpha_i}(x), \qquad (4.9)$$

here $g_{a,\alpha_i} : X \to X$. This scheme is applicable both in classical and quantum physics as well as outside of it, e.g., in psychology Stimulus-Organism-Response (S-O-R) scheme for explaining behavior, see, e.g., [166], of humans and other cognitive systems, see Chapter 9. Measurement of a is performed by some physical *apparatus* $\mathcal{A}$. Its statistical properties are determined by a pair of maps $(f_{a,\alpha}, g_{a,\alpha})$. In fact, each such pair represents a class of physical apparatuses having the same statistical properties.

If a state space X can be embedded into a linear space $Y : X \subset Y$, then the map

$$\mathcal{E}(\alpha, x) = f_{a,\alpha}(x)g_{a,\alpha}(x) \qquad (4.10)$$

is called *instrument*. (The same instrument represents a class of physical apparatuses.) We remark that, for each value α, the map $x \to \mathcal{E}(\alpha, x)$ is from X to Y. Denote it by $\mathcal{E}(\alpha)$, i.e., $\mathcal{E}(\alpha)x = \mathcal{E}(\alpha, x)$.

Now turn to QM. The specialty of the theory of quantum instruments is that the state space X is selected as the space of density operators in a complex Hilbert space, $X = D(H)$ and $Y = L(H)$ is the space of all linear operators in H (in the infinite dimensional case we consider the space of linear bounded operators). Here, for $\rho \in D(H)$,

$$p(\alpha|\rho) = f_{a,\alpha}(\rho) = \mathrm{Tr}\mathcal{E}(\alpha)\rho \qquad (4.11)$$

and

$$\rho_\alpha = g_{a,\alpha}(\rho) = \frac{\mathcal{E}(\alpha)\rho}{\mathrm{Tr}\mathcal{E}(\alpha)\rho}, \qquad (4.12)$$

see sections 4.5, 4.12 for details. The trace rule (4.11) for probability is a consequence of the Born rule.

4.2.1 *Projection measurements*

We shall use Latin letters $a, b, \ldots$ to denote physical observables and the corresponding capital letters $A, B, \ldots$ to denote the corresponding operational representations of these observables (typically Hermitian operators) in the quantum formalism. Such distinguishing is useful for further considerations, see Chapter 5, where we shall consider general contextual probability approach to observations.

We start with the standard von Neumann-Lüders measurements, which gives us an important class of quantum instruments (especially from the historical viewpoint). For this class of measurements, observables are mathematically represented by *Hermitian operators*,

$$A = \sum_i \alpha_i P_{\alpha_i}, \qquad (4.13)$$

where P_{α_i} is the projector onto the eigensubspace corresponding to the eigenvalue α_i belonging to $\mathbf{R}$.

As was already mentioned, by the axiomatics of QM *the possible results of measurements are represented by values α_i belonging to the operator's spectrum.* Since we proceed in the finite-dimensional case, the spectrum coincides with the set of eigenvalues.

For pure states, the transformation (4.12) is based on the projection P_{α_i} :

$$\psi \to P_{\alpha_i}\psi, \qquad (4.14)$$

this map is linear and convenient to work with. However, if $P_{\alpha_i} \neq I$, where I is the unit operator, then $\|P_{\alpha_i}\psi\| < 1$, so the output of (4.14) is not a state. To obtain a state, we have to normalized the output its norm:

$$\psi \to \frac{P_{\alpha_i}\psi}{\|P_{\alpha_i}\psi\|}. \qquad (4.15)$$

This is a map from the space of pure states into the space of pure states. Note that it is a nonlinear map.

As was pointed out in section 4.1, for a pure state ψ, one can consider its representation by the density operator $\rho = P_\psi$, the projector on the vector ψ, see (4.4). In such terms, the state transform (4.14) can be written as

$$\rho \to P_{\alpha_i}\rho P_{\alpha_i}. \qquad (4.16)$$

This is the simplest example of a transformation which in the quantum measurement theory is called a *quantum operation.* It can be extended to a linear map from the space of linear operators (matrices) to itself – by the same formula (4.16). For a finite spectral set O, the collection of quantum operations (4.16), $\alpha_i \in O$, gives the simplest example of a *quantum instrument:* $\mathcal{E}(\alpha)\rho = P_\alpha\rho P_\alpha$.

We are interested in the corresponding map from the space of density operators (matrices) to itself, see (4.12). Thus we again have to make normalization:

$$\rho \to \rho_{\alpha_i} = \frac{P_{\alpha_i}\rho P_{\alpha_i}}{\mathrm{Tr}P_{\alpha_i}\rho P_{\alpha_i}} = \frac{\mathcal{E}(\alpha)\rho}{\mathrm{Tr}\mathcal{E}(\alpha)\rho}. \qquad (4.17)$$

It is nonlinear and physicists work with quantum operations (forming instruments), by making normalization by trace only at the final step of calculations which can involve a chain of measurements.

We are interested not only in the measurement impact on the initial quantum state ρ (represented by a quantum instrument), but also in the probabilities of the observed results $\alpha_i \in O, p(\alpha_i|\rho)$. They are given by *the Born rule,* the basic rule coupling quantum theory with experimental statistics.[4] If the initial state is pure $\rho = P_\psi$, then the Born rule has the form:

$$p(\alpha_i|\psi) = \langle P_{\alpha_i}\psi|\psi\rangle = \|P_{\alpha_i}\psi\|^2. \tag{4.18}$$

It is easy to see that

$$p(\alpha_i|\psi) = \text{Tr } P_{\alpha_i}P_\psi. \tag{4.19}$$

This formula can be easily generalized, e.g., via (4.5), to an arbitrary initial state ρ :

$$p(\alpha_i|\rho) = \text{Tr } P_{\alpha_i}\rho = \text{Tr } \mathcal{E}(\alpha)\rho. \tag{4.20}$$

In the von Neumann-Lüders approach the quantum instrument is uniquely determined by the observable, represented mathematically by the Hermitian operator A. The latter is the basis of the construction. However, even in this approach we could start directly with an instrument determined by a family of mutually orthogonal projectors (P_{α_i}), i.e.,

$$\sum_i P_{\alpha_i} = I, \tag{4.21}$$

where $P_{\alpha_i} \perp P_{\alpha_j}, i \neq j$, and then define the quantum observable A simply as this family (P_{α_i}). In quantum information the values α_i have merely the meaning of *labels* for the results of measurement. For future generalization, we remark that the normalization condition (4.21) can be written as

$$\sum_i P_{\alpha_i}^\star P_{\alpha_i} = I, \tag{4.22}$$

because, for any orthogonal projector P, $P^\star = P$ and $P^2 = P$.

4.2.2 *Projection postulate for pure states*

Consider a pure state ψ and a quantum observable A of the von Neumann-Lüders type (representing some physical observable a). Suppose that the Hermitian operator A has non-degenerate spectrum; denote its eigenvalues by $\alpha_1, ..., \alpha_m$ and the corresponding eigenvectors by $e_1, ..., e_m$ (here $\alpha_i \neq \alpha_j, i \neq j$). This is an orthonormal basis in H. We expand the vector ψ with respect to this basis:

$$\psi = k_1 e_1 + ... + k_m e_m, \tag{4.23}$$

---

[4]This is a postulate. This rule cannot be derived with the aid of quantum theory. It is justified by its applicability to quantum phenomena. See [167], [256] for possible deviations from the Born rule; see also [167] for an attempt to derive the Born rule from a subquantum model of the classical random field type.

where (k_j) are complex numbers such that

$$\|\psi\|^2 = |k_1|^2 + \ldots + |k_m|^2 = 1. \qquad (4.24)$$

By using the terminology of linear algebra we say that the pure state ψ is a *superposition* of the pure states e_j.

For the pure state ψ given by (4.23), the Born rule (4.18) has the form

$$p(\alpha_j) = |k_j|^2 = |\langle \psi | e_j \rangle|^2. \qquad (4.25)$$

We remark that $\sum_j p(\alpha_j) = \|\psi\|^2 = 1$. State normalization simply encodes normalization of the probability distributions for all quantum observables given by Hermitian operators, i.e., of the von Neumann-Lüders type.

In this case (of an observable with nondegenerate spectrum and a pure state) the projection postulate can be formulated as follows:

Measurement of an observable a (represented by the Hermitian operator A) resulting in α_i, induces reduction ("collapse") of the superposition (4.23) to the basis vector e_i corresponding to this eigenvalue α_i.

This procedure can be interpreted in the following way:

Superposition (4.23) reflects uncertainty in the results of measurements for an observable a. Before measurement a quantum system "does not know how it will answer to the question a." The mathematical expression (4.25) encodes potentialities for different answers. Thus a quantum system in the superposition state ψ does not have propensity to any value of a as its objective property. After the measurement, the superposition is reduced to the single term in the expansion (4.23) corresponding to the value of a obtained in the process of measurement.

The projection postulate is one of the most questionable and debatable postulates of QM, and we postpone the interpretational discussion to section 6.10.

4.3 Conditional Probabilities

There are given a state represented by the density operator ρ and an observable a represented by the Hermitian operator A with the spectral decomposition (4.13): $A = \sum_i \alpha_i P_{\alpha_i}$. The probability to obtain a concrete value $a = \alpha_i$ as the result of measurement of the observable a is given by the Born rule (4.20).

Suppose now that after measurement of the a-observable one plans to perform measurement of another observable b represented by the Hermitian operator B with the spectral decomposition $B = \sum_i \beta_i Q_{\beta_i}$, where Q_{β_i} are orthogonal projectors onto the eigensubspaces corresponding to the eigen-values β_i. Then to predict the probability of the result $b = \beta_i$, one can use the output state ρ_{α_i}. This state is determined by the projection postulate, see (4.17). For the b-measurement following the a-measurement, this state, ρ_{α_i}, plays the same role as the state ρ played for the preceding a-measurement. In particular, by applying the Born rule once again we obtain:

$$p(\beta_j|\rho_{\alpha_i}) = \text{Tr}Q_{\beta_j}\rho_{\alpha_i} = \frac{\text{Tr}Q_{\beta_j}P_{\alpha_i}\rho P_{\alpha_i}}{\text{Tr}P_{\alpha_i}\rho P_{\alpha_i}}. \tag{4.26}$$

In QM, especially in quantum information theory, this probability is often called the *conditional probability*[5]

$$P_\rho(b = \beta_j|a = \alpha_i) \equiv p(\beta_j|\rho_{\alpha_i}). \tag{4.27}$$

The rule (4.26) can be considered as a quantum analog of the Bayes formula for classical conditional probability. For any fixed output $a = \alpha_i$, the following probability normalization condition holds:

$$\sum_\beta P_\rho(b = \beta|a = \alpha_i) = 1. \tag{4.28}$$

Consider now the case of successive measurements of the same observable a. Here we have

$$P_\rho(a = \alpha_j|a = \alpha_i) = \frac{\text{Tr}P_{\alpha_j}P_{\alpha_i}\rho P_{\alpha_i}}{\text{Tr}P_{\alpha_i}\rho P_{\alpha_i}} = \delta_{ij}. \tag{4.29}$$

Thus if the result $a = \alpha_i$ was registered in the first measurement of a, then in its next measurement the same result will be registered with probability 1. This is the *repeatability* feature of the observables of the von Neumann-Lüders type. In this sense they are natural generalizations of classical random observables, see section 1.5.

Suppose that both operators A and B have non-degenerate spectra, i.e., each eigensubspace is one dimensional. Consider two orthonormal bases

---

[5]This terminology has some degree of ambiguity. The meaning of such probability is not precisely the same as the meaning of classical conditional probability. In QM we do not consider conditioning with respect to some event, but rather with respect to the state ρ_{α_i} which is the output of the earlier measurement. Later, by considering observables with nondegenerate spectra we shall invent the terminology "transition probabilities". This terminology is adequate to this very special case. However, in the case of observables with degnerate spectra it is also ambiguous. My personal viewpoint is that the probabilities under consideration has to be treated as contextual probabilities. Here measurement of a with the fixed output $a = \alpha$ determines the context for next measurement. General probabilistic contextual model of measurements (of any type: classical, quantum; performed in physics, biology, psychology, sociology) will be presented in section 5.5.

composed of eigenvectors of these operators, $\{f_{\alpha_i}\}, \{e_{\beta_i}\}$. We also suppose that the initial input state is a pure state ψ. Then a-measurement with the result $a = \alpha$ induces projection of the state-vector ψ onto the eigenvector f_α. This is the essence of the projection postulate. Thus, for consequent measurement of b, this eigenvector f_α of A plays the role of the input state. Then, as the result of b-measurement with the output $b = \beta$, the state is projected onto e_β. Here conditional probabilities given by (4.27) do not depend on the initial state ψ :

$$P(b = \beta_j | a = \alpha_i) = |\langle f_\alpha | e_\beta \rangle|^2. \tag{4.30}$$

We introduce the matrix of *transition probabilities* $\mathbf{P}^{b|a} = (p_{\beta|\alpha})$, which elements $p_{\beta|\alpha}$ are determined as conditional probabilities given by (4.30). Here $p_{\beta|\alpha}$ is the probability of transitions from the state f_α determined by the output α of a-measurement to the state e_β determined by the output β of b-measurement.

We remark that, for each f_α,

$$\sum_\beta |\langle f_\alpha | e_\beta \rangle|^2 = \|f_\alpha\|^2 = 1. \tag{4.31}$$

Therefore, for each α belonging to the range of values of a (the spectrum of A), we have:

$$\sum_\beta p_{\beta|\alpha} = 1. \tag{4.32}$$

We recall that the matrix $\mathbf{P}^{b|a}$ is called *stochastic*. We also have, for each fixed β,

$$\sum_\alpha |\langle f_\alpha | e_\beta \rangle|^2 = \|e_\beta\|^2 = 1. \tag{4.33}$$

Thus, for each β from the range of values of b (the spectrum of B), we have:

$$\sum_\alpha p_{\beta|\alpha} = 1. \tag{4.34}$$

We recall that the matrix $\mathbf{P}^{b|a}$ with elements satisfying both constraints (4.32), (4.34) is called *doubly stochastic*.

Hence, the matrix of transition probabilities generated by any pair of observables (of the von Neumann-Lüders type) having nondegenerate spectra is doubly stochastic, cf. section 5.4.5.

The symmetry of the scalar product also implies that the elements of the matrices of transition probabilities from the a-states to the b-states and vice versa, namely, $\mathbf{P}^{b|a}$ and $\mathbf{P}^{a|b}$, are coupled by the symmetry relation:

$$p_{\beta|\alpha} = p_{\alpha|\beta}. \tag{4.35}$$

Such observables we call *symmetrically conditioned*. This is an important feature of pairs of observables (of the von Neumann-Lüders type) having nondegenerate spectra, i.e., one dimensional eigensubspaces. In section 5.4.5 we shall consider its generalization to general contextual random measurements (and quantum measurements will be treated as the special subclass of such measurements).

We started with a pure state ψ only for illustrative reasons. In the same way we can start with any density operator ρ. If A has a nondegnerate spectrum, then $\rho_\alpha = P_\alpha$ is the projector on the vector f_α, i.e., this is a pure state. Hence, finally we shall obtain the same formula (4.30) with all its consequences, in particular, double stochasticity of the matrix of transition probabilities.

We remark that the matrix of "transition probabilities" $\mathbf{P}^{b|a}$ can be defined even in the case of observables with degenerate spectra, but here its elements depend on the initial input state, see (4.27). This matrix is also stochastic. However, in general it is not doubly stochastic.

4.4 Quantum Logic

Von Neumann and Birkhoff [250], [35] suggested to represent *events* (propositions) by orthogonal projectors in complex Hilbert space H.

For an orthogonal projector P, we set $H_P = P(H)$, its image, and vice versa, for subspace L of H, the corresponding orthogonal projector is denoted by the symbol P_L.

The set of orthogonal projectors is a *lattice* with the order structure: $P \leq Q$ iff $H_P \subset H_Q$ or equivalently, for any $\psi \in H$, $\langle \psi | P \psi \rangle \leq \langle \psi | Q \psi \rangle$.

We recall that the lattice of projectors is endowed with operations "and" ($\wedge$) and "or" ($\vee$). For two projectors P_1, P_2, the projector $R = P_1 \wedge P_2$ is defined as the projector onto the subspace $H_R = H_{P_1} \cap H_{P_2}$ and the projector $S = P_1 \vee P_2$ is defined as the projector onto the subspace H_R defined as the minimal linear subspace containing the set-theoretic union $H_{P_1} \cup H_{P_2}$ of subspaces H_{P_1}, H_{P_2} : this is the space of all linear combinations of vectors belonging to these subspaces. The operation of negation is defined as the orthogonal complement: $P^{\perp} = \{y \in H : \langle y | x \rangle = 0 \text{ for all } x \in H_P\}$.

In the language of subspaces the operation "and" coincides with the usual set-theoretic intersection, but the operations "or" and "not" are nontrivial deformations of the corresponding set-theoretic operations. It is natural to expect that such deformations can induce deviations from classical Boolean logic.

Consider the following simple example. Let H be two dimensional Hilbert space with the orthonormal basis (e_1, e_2) and let $v = (e_1 + e_2)/\sqrt{2}$. Then $P_v \wedge P_{e_1} = 0$ and $P_v \wedge P_{e_2} = 0$, but $P_v \wedge (P_{e_1} \vee P_{e_2}) = P_v$. Hence, for quantum events, in general the distributivity law is violated:

$$P \wedge (P_1 \vee P_2) \neq (P \wedge P_1) \vee (P \wedge P_2). \tag{4.36}$$

The lattice of orthogonal projectors is called *quantum logic*. It is considered as a (very special) generalization of classical Boolean logic. Any sub-lattice consisting of commuting projectors can be treated as classical Boolean logic.

At the first sight the representation of events by projectors/linear subspaces might look exotic. However, this is simply a prejudice which springs from too common usage of the set-theoretic representation of events (Boolean logic) in the modern classical probability theory. The tradition to represent events by subsets was firmly established by A. N. Kolmogorov in 1933. We remark that before him the basic classical probabilistic models were not of the set-theoretic nature. For example, the main competitor of the Kolmogorov model, the von Mises frequency model, was based on the notion of a collective.

As we have seen, quantum logic relaxes some constraints posed on the operations of classical Boolean logic, in particular, the distributivity constraint. This provides novel possibilities for logically consistent reasoning, cf. with the information interpretation of QM (see section 6.5) and applications of the mathematical formalism of QM (in fact, reasoning and decision making based on quantum logic) outside of physics (Chapter 9).

4.5   Atomic Instruments

Now we move to theory of *atomic instruments*. Here quantum operations have the form:

$$\rho \to Q_{\alpha_i} \rho Q_{\alpha_i}^\star, \tag{4.37}$$

where, for each value α_i, Q_{α_i} is a linear operator which is a contraction (i.e., its norm is bounded by 1). These operators are constrained by the normalization condition, cf. (4.22):

$$\sum_i M_{\alpha_i} = I, \quad \text{where } M_{\alpha_i} = Q_{\alpha_i}^\star Q_{\alpha_i}. \tag{4.38}$$

The quantum operations (4.37) determine an *atomic quantum instrument* $\mathcal{E}(\alpha)\rho = Q_\alpha \rho Q_\alpha^\star$ which induces the corresponding state transformation:

$$\rho \to \rho_{\alpha_i} = \frac{Q_{\alpha_i} \rho Q_{\alpha_i}^\star}{\text{Tr}\, Q_{\alpha_i} \rho Q_{\alpha_i}^\star} = \frac{\mathcal{E}(\alpha)\rho}{\text{Tr}\mathcal{E}(\alpha)\rho}. \tag{4.39}$$

In particular, pure states are transformed into pure states (similar to the von Neumann-Lüders instruments):

$$\psi \to \frac{Q_{\alpha_i}\psi}{\|Q_{\alpha_i}\psi\|}. \tag{4.40}$$

Now it is time to define generalized quantum observables:

Definition. *A positive operator valued measure (POVM) is a family of positive-semidefinite operators* $\{M_j\}$ *such that* $\sum_{j=1}^{m} M_j = I$, *where* I *is the unit operator.*

Thus the family of operators $\{M_{\alpha_i}\}$, see (4.38), is POVM. Probabilities of the results of measurements are given by the following extension of the Born rule (4.20) on POVMs :

$$p(\alpha_i|\rho) = \mathrm{Tr}\, M_{\alpha_i}\rho, \tag{4.41}$$

where

$$M_{\alpha_i} = Q_{\alpha_i}^{\star} Q_{\alpha_i} \tag{4.42}$$

or by using the instrument notation:

$$p(\alpha_i|\rho) = \mathrm{Tr}\mathcal{E}(\alpha)\rho.$$

(We remark that if Q_{α_i} is a projector, then $Q_{\alpha_i}^{\star} = Q_{\alpha_i}$ and $Q_{\alpha_i}^{2} = Q_{\alpha_i}$. Thus in this case (4.41) matches (4.20).) The class of atomic instruments is the straightforward generalization of the von Neumann-Lüders class. In general, quantum instruments do not transfer pure states into pure states.

In the same way as in the von Neumann-Lüders case we can define conditional probability: by using the state transform (4.39). One of the main differences is that atomic instruments do not have the property of repeatability. From getting some value α_i in the first measurement we cannot confidently conclude that the same value will be registered with probability one in successive measurement of the same observable.

4.6 Symmetric Informationally Complete Quantum Instruments

We consider one special class of atomic instruments with quantum observables given by *symmetric informationally complete* POVMs, SIC-POVMs. Here informational completeness means that the probabilities of observing the various outcomes (given by Born's rule) entirely determine any quantum state ρ being measured. This requires d^2 linearly independent operators for the state space of the dimension d.

The simplest definition is that a SIC-POVM is determined by a system of d^2 normalized vectors (ϕ_i) (they are not orthogonal) such that

$$\langle\phi_i|\phi_j\rangle^2 = \frac{1}{d+1}, i \neq j. \tag{4.43}$$

The elements of the corresponding SIC-POVM (E_i) are subnormalized projectors $E_i = \frac{1}{d}\Pi_i$, where Π_i is the orthogonal projector on ϕ_i. The elements of SIC-POVM E_i determine the corresponding quantum operations (atomic instruments).

The characteristic property of SIC-POVMs, symmetry, is that the inner product in the space of operators (or $d \times d$ matrices) given by the trace is constant, i.e.,

$$\mathrm{Tr}E_iE_j = \mathrm{const} = \frac{1}{d^2(d+1)}, i \neq j.$$

By using this equality it is easy to obtain the following representation of an arbitrary density operator ρ :

$$\rho = \sum_i \left((d+1)p(i) - \frac{1}{d}\right)\Pi_i \tag{4.44}$$

where $p(i) = \mathrm{Tr}E_i\rho$ is the probability to obtain the result i for a measurement presented by the SIC-POVM (E_i).

This SIC-POVM based representation of a density operator ρ plays the crucial role in Quantum Bayesianism (QBism), cf. with the QBist version of quantum generalization of FTP (see section 6.7.3).

4.7 Schrödinger and von Neumann Equations

In QM the state dynamics of an isolated quantum system is described by Schrödinger's equation:

$$i\hbar\frac{d\psi}{dt} = \mathcal{H}\psi(t), \qquad \psi(0) = \psi^0, \tag{4.45}$$

where $\mathcal{H}$ is a Hermitian positive-semidefinite operator, Hamiltonian, representing system's energy, and $\hbar = h/2\pi$ is the reduced Planck constant.

The Schrödinger equation is, in fact, a system of linear differential equations with complex coefficients; in the one dimensional case $\mathcal{H}$ is just a real number and the general solution has the form of the imaginary exponent: $\psi(t) = e^{\frac{-it\mathcal{H}}{\hbar}}\psi^0, \mathcal{H} \in \mathcal{R}$. In the general case $\mathcal{H}$ is an operator and the solution is represented in the form of imaginary operator-exponent (for the fixed basis it is simply the exponent of the matrix):

$$\psi(t) = U_t\psi^0, \; U_t = e^{\frac{-it\mathcal{H}}{\hbar}}. \tag{4.46}$$

As well as the one dimensional imaginary exponent, the operator-exponent describes *oscillating dynamics*. It is more complicated than in the one dimensional case; it is a mixture of many oscillating imaginary exponents.

We point to the following fundamental property of the Schrödinger dynamics. The evolution operator U_t, see (4.46), is a *unitary operator*. It preserves the scalar product:

$$\langle U_t \psi_1 | U_t \psi_2 \rangle = \langle \psi_1 | \psi_2 \rangle \tag{4.47}$$

and its inverse exists and is given as $U_t^{-1} = U_t^\star$.

This dynamics transfers a pure quantum state to another pure quantum state. This preservation of the state norm physically means conservation of total probability.

Since any general quantum state, a density operator ρ, can be represented as a mixture of density operators corresponding to pure states, see (4.5), the Schrödinger dynamics for pure states implies the following dynamics for density operators:

$$\gamma \frac{d\rho}{dt}(t) = -\frac{i}{\hbar}[H, \rho(t)], \qquad \rho(0) = \rho^0. \tag{4.48}$$

This equation is known as the *von Neumann equation* [250]. By using representation (4.46) of the Schrödinger evolution for the pure state we represent the evolution of the density operator in the form

$$\rho(t) = U_t^* \rho^0 U_t. \tag{4.49}$$

4.8 Compound Systems

The quantum description of a compound system S consisting of two subsystems S_1 and S_2 with state (Hilbert) spaces H_1 and H_2, respectively, is based on the representation of the states of this system in the tensor product space $H = H_1 \otimes H_2$ (the definition can be found below).

Observables (of the von Neumann-Lüders type) on compound systems are represented as Hermitian operators acting in $H = H_1 \otimes H_2$. An important class of observables on compound systems consists of *local observables*: there are two observables a_1 and a_2 on S_1 and S_2, respectively. Their joint measurement, i.e., measurement of the pair (a_1, a_2), can be treated as measurement on S. If mathematically the observables a_i are represented by the Hermitian operators $A_i, i = 1, 2$, acting in H_i, then the corresponding measurement on S is represented as the tensor product of the Hermitian operators representing the local observables: $A = A_1 \otimes A_2$ (the definition of such tensor product can be found below).

Since the notion of *tensor product* of Hilbert spaces is not used so much in classical probability theory[6], we shall present briefly the construction of Hilbert space $H_1 \otimes H_2$.

Although in quantum information theory and in this book we shall use the formal algebraic definition, which is especially useful for finite dimensional Hilbert spaces, we prefer to start with the construction originally used by von Neumann [250], namely, the tensor product of two spaces of square integrable functions. These Hilbert spaces are infinite dimensional, but they have the functional space structure which makes them easily handleable by readers.

Tensor product of functional spaces. Let both state spaces be L^2-spaces, $H_1 = L^2(\mathbb{R}^k)$ and $L^2(\mathbb{R}^m)$.

Take two functions; $\psi \equiv \psi(x)$ belongs to H_1 and $\phi \equiv \phi(y)$ belongs to H_2. By multiplying these functions we obtain the function of two variables $\Psi(x,y) = \psi(x) \times \phi(y)$. It is easy to check that this function belongs to the space $H = L^2(\mathbb{R}^{k+m})$. Take now n functions, $\psi_1(x), ..., \psi_n(x)$, from H_1 and n functions, $\phi_1(y), ..., \phi_n(y)$, from H_2 and consider the sum of their pairwise products:

$$\Psi(x,y) = \sum_i \psi_i(x) \times \phi_i(y). \tag{4.50}$$

This function also belongs to H.

It is possible to show that any function belonging to H can be represented as (4.50), where the sum is in general infinite. Multiplication of functions is the basic example of the operation of the tensor product. The latter is denoted by the symbol $\otimes$. Thus in the example under consideration $\psi \otimes \phi(x,y) = \psi(x) \times \phi(y)$. The tensor product structure on $H = L^2(\mathbb{R}^{k+m})$ is symbolically denoted as $H = H_1 \otimes H_2$.

Consider now orthonormal bases in H_k, $(e_j^{(k)})$, $k = 1, 2$. Then (functions) $(e_{ij} = e_i^{(1)} \otimes e_j^{(2)})$ form an orthonormal basis in H : any $\Psi \in H$, can be represented as

$$\Psi = \sum c_{ij} e_{ij} \equiv \sum c_{ij} e_i^{(1)} \otimes e_j^{(2)}, \tag{4.51}$$

where

$$\sum |c_{ij}|^2 < \infty. \tag{4.52}$$

Those who work with electromagnetic signals have experience in expanding electromagnetic signals with respect to various bases, e.g., using

---

[6] In the coordinate form tensor products of vectors and matrices are also known under the name *Kronecker product*.

the Fourier expansion or the wavelet expansion. Some bases are indexed by continuous parameters; integrals take place of sums. Thus the notion of basis in the L^2-space is widely known.

However, there is a crucial difference between the classical field and quantum mechanical representations of compound systems. The state of a classical bi-signal consisting of two components is represented in the *Cartesian product* of the corresponding L^2-spaces. And the state of a quantum bi-system, e.g., bi-photon, is represented in the tensor product space. One may state that the crucial difference between the classical and quantum physical models is in the representation of states of compound systems[7], cf., however, with the classical random field representation of compound quantum systems [167] (in the framework of *prequantum classical statistical field theory*).

Tensor product, the algebraic definition. Consider now two finite dimensional Hilbert spaces, H_1, H_2. For each pair of vectors $\psi \in H_1, \phi \in H_2$, we form a new formal entity denoted by $\psi \otimes \phi$. Then we consider the sums $\Psi = \sum_i \psi_i \otimes \phi_i$. On the set of such formal sums we can introduce the linear space structure. (To be mathematically rigorous, we have to constraint this set by some algebraic relations to make the operations of addition and multiplication by complex numbers well defined.) This construction gives us the tensor product $H = H_1 \otimes H_2$. In particular, if we take orthonormal bases in $H_k, (e_j^{(k)}), k = 1, 2$, then $(e_{ij} = e_i^{(1)} \otimes e_j^{(2)})$ form an orthonormal basis in H, any $\Psi \in H$, can be represented as (4.51) with (4.52).

The latter representation gives the simplest possibility to define the tensor product of two arbitrary (i.e., may be infinite-dimensional) Hilbert spaces as the space of formal series (4.51) satisfying the condition (4.52).

Besides the notion of the tensor product of states, we shall also use the notion of the tensor product of operators. Consider two linear operators $A_i : H_i \to H_i, i = 1, 2$. Their tensor product $A \equiv A_1 \otimes A_2 : H \to H$ is defined starting with the tensor products of two vectors:

$$A\psi \otimes \phi = (A_1 \psi) \otimes (A_2 \phi).$$

Then it is extended by linearity. By using the coordinate representation (4.51) the tensor product of operators can be represented as

$$A\Psi = \sum c_{ij} A e_{ij} \equiv \sum c_{ij} A_1 e_i^{(1)} \otimes A_2 e_j^{(2)}. \tag{4.53}$$

---

[7]Although, as we have already seen in the previous sections, the descriptions of non-compound systems differ essentially.

If the operators $A_i, i = 1, 2$, are represented by matrices (with respect to the fixed bases), i.e., $A_i = (A_{kl}^{(i)})$, then the matrix $A = (A_{kl.nm})$ with respect to the tensor product of these bases can be easily calculated.

In the same way one defines the tensor product of Hilbert spaces, $H_1, ..., H_n$, denoted by the symbol $H = H_1 \otimes ... \otimes H_n$. We start with forming the formal entities $\psi_1 \otimes ... \otimes \psi_n$, where $\psi_j \in H_j, j = 1, ..., n$. Tensor product space is defined as the set of all sums $\sum_j \psi_{1j} \otimes ... \otimes \psi_{nj}$ (which has to be constrained by some algebraic relations, but we omit such details). Take orthonormal bases in $H_k, (e_j^{(k)}), k = 1, ..., n$. Then any $\Psi \in H$ can be represented as

$$\Psi = \sum_\alpha c_\alpha e_\alpha \equiv \sum_{\alpha = (j_1 ... j_n)} c_{j_1 ... j_n} e_{j_1}^{(1)} \otimes ... \otimes e_{j_n}^{(n)}, \qquad (4.54)$$

where $\sum_\alpha |c_\alpha|^2 < \infty$.

4.9   Dirac's Symbolic Notations

Dirac's notations [79] are widely used in quantum information theory. Vectors of H are called *ket-vectors*, they are denoted as $|\psi\rangle$. The elements of the dual space H' of H, the space of linear continuous functionals on H, are called *bra-vectors*, they are denoted as $\langle\psi|$.

Originally the expression $\langle\psi|\phi\rangle$ was used by Dirac for the duality form between H' and H, i.e., $\langle\psi|\phi\rangle$ is the result of application of the linear functional $\langle\psi|$ to the vector $|\phi\rangle$. In mathematical notation it can be written as follows. Denote the functional $\langle\psi|$ by f and the vector $|\phi\rangle$ by simply ϕ. Then $\langle\psi|\phi\rangle \equiv f(\phi)$. To simplify the model, later Dirac took the assumption that H is Hilbert space, i.e., the H' can be identified with H. We remark that this assumption is an axiom simplifying the mathematical model of QM. However, in principle Dirac's formalism [79] is applicable for any topological linear space H and its dual space H'; so it is more general than von Neumann's formalism [250] rigidly based on Hilbert space.

Consider an observable a given by the Hermitian operator A with nondegenerate spectrum and restrict our consideration to the case of finite dimensional H. Thus the normalized eigenvectors e_i of A form the orthonormal basis in H. Let $Ae_i = \alpha_i e_i$. In Dirac's notation e_i is written as $|\alpha_i\rangle$ and, hence, any pure state can be written as

$$|\psi\rangle = \sum_i c_i |\alpha_i\rangle, \quad \sum_i |c_i|^2 = 1. \qquad (4.55)$$

Since the projector onto $|\alpha_i\rangle$ is denoted as $P_{\alpha_i} = |\alpha_i\rangle\langle\alpha_i|$, the operator A can be written as

$$A = \sum_i \alpha_i |\alpha_i\rangle\langle\alpha_i|. \tag{4.56}$$

Now consider two Hilbert spaces H_1 and H_2 and their tensor product $H = H_1 \otimes H_2$. Let $(|\alpha_i\rangle)$ and $(|\beta_i\rangle)$ be orthonormal bases in H_1 and H_2 corresponding to the eigenvalues of two observables A and B. Then vectors $|\alpha_i\rangle \otimes |\beta_j\rangle$ form the orthonormal basis in H. Typically in physics the sign of the tensor product is omitted and these vectors are written as $|\alpha_i\rangle|\beta_j\rangle$ or even as $|\alpha_i\beta_j\rangle$. Thus any vector $\psi \in H = H_1 \otimes H_2$ can be represented as

$$\psi = \sum_{ij} c_{ij} |\alpha_i\beta_j\rangle, \tag{4.57}$$

where $c_{ij} \in \mathbf{C}$ (in the infinite-dimensional case these coefficients are constrained by the condition $\sum_{ij} |c_{ij}|^2 < \infty$).

4.10 Quantum Bits

In particular, in quantum information theory typically qubit states are represented with the aid of observables having the eigenvalues $0, 1$. Each qubit space is two dimensional:

$$|\psi\rangle = c_0|0\rangle + c_1|1\rangle, \ |c_0|^2 + |c_1|^2 = 1. \tag{4.58}$$

A pair of qubits is represented in the tensor product of single qubit spaces, here pure states can be represented as superpositions:

$$|\psi\rangle = c_{00}|00\rangle + c_{01}|01\rangle + c_{10}|10\rangle + c_{11}|00\rangle, \tag{4.59}$$

where $\sum_{ij} |c_{ij}|^2 = 1$. In the same way the n-qubit state is represented in the tensor product of n one-qubit state spaces (it has the dimension 2^n) :

$$|\psi\rangle = \sum_{x_j=0,1} c_{x_1...x_n} |x_1...x_n\rangle, \tag{4.60}$$

where $\sum_{x_j=0,1} |c_{x_1...x_n}|^2 = 1$. We remark that the dimension of the n qubit state space grows exponentially with the growth of n. The natural question about possible physical realizations of such multi-dimensional state spaces arises. The answer to it is not completely clear; it depends very much on the used interpretation of the wave function (see Chapter 6).

4.11   Entanglement

Consider the tensor product $H = H_1 \otimes H_2 \otimes ... \otimes H_n$ of Hilbert spaces $H_k, k = 1, 2, ..., n$. The states of the space H can be *separable and non-separable* (entangled). We start by considering pure states. The states from the first class, separable pure states, can be represented in the form:

$$|\psi\rangle = \otimes_{k=1}^{n}|\psi_k\rangle = |\psi_1...\psi_n\rangle, \qquad (4.61)$$

where $|\psi_k\rangle \in H_k$. The states which cannot be represented in this way are called non-separable, or entangled. Thus mathematically the notion of entanglement is very simple, it means impossibility of tensor factorization.

For example, let us consider the tensor product of two one-qubit spaces. Select in each of them an orthonormal basis denoted as $|0\rangle, |1\rangle$. The corresponding orthonormal basis in the tensor product has the form $|00\rangle, |01\rangle, |10\rangle, |11\rangle$. Here we used Dirac's notations, see section 4.9, near the end. Then so called Bell's states

$$|\Phi^+\rangle = (|00\rangle + |11\rangle)/\sqrt{2}, \quad |\Phi^-\rangle = (|00\rangle - |11\rangle)/\sqrt{2}; \qquad (4.62)$$

$$|\Psi^+\rangle = (|01\rangle + |10\rangle)/\sqrt{2}, \quad |\Psi^-\rangle = (|01\rangle - |10\rangle)/\sqrt{2}, \qquad (4.63)$$

are entangled.

Although the notion of entanglement is mathematically simple, its physical interpretation is one of the main problems of modern quantum foundations. The common interpretation is that entanglement encodes quantum nonlocality, the possibility of action at the distance (between parts of a system in an entangled state). Such an interpretation implies the drastic change of all classical physical presentations about nature, at least about the microworld. In the probabilistic terms entanglement induces correlations which are too strong to be described by classical probability theory. (At least this is the common opinion of experts in quantum information theory, cf. Chapter 6, see also [161].) Such correlations violate the famous Bell inequality (Chapter 5) which can be derived only in classical probability framework. The latter based on the use of a single probability space covering probabilistic data collected in a few incompatible measurement contexts, see section 1.3.2.

Now consider a quantum state given by a density operator in H. This state is called separable if it can be factorized in the product of density operators in spaces H_k :

$$\rho = \otimes_{k=1}^{n}\rho_k, \qquad (4.64)$$

otherwise the state ρ is called entangled. We remark that an interpretation of entanglement for mixed states is even more complicated than for pure states.

4.12 General Theory of Quantum Instruments

We recall that the symbol $D(H)$ denotes the space of density operators in the complex Hilbert space H; $L(H)$ the space of all linear operators in H (bounded operators in the infinite dimensional case). The space $L(H)$ can itself be endowed with the structure of the linear space. We also have to consider linear operators from $L(H)$ into itself; such maps, $T : L(H) \to L(H)$ are called *superoperators*. We shall use this notion only in section 4.12.1. Thus, for the moment, the reader can proceed without it.

Moreover, on the space $L(H)$ it is possible to introduce the structure of Hilbert space with the scalar product

$$\langle A|B \rangle = \operatorname{Tr} A^* B. \tag{4.65}$$

Therefore, for each superoperator $T : L(H) \to L(H)$, there is defined its adjoint (super)operator $T^* : L(H) \to L(H)$, $\langle T(A)|B \rangle = \langle A|T^*(B) \rangle$, $A, B \in L(H)$.

Consider measurement of an observable a having the discrete set of outputs $O = \{\alpha_1, ..., \alpha_n\}$. A physical apparatus $\mathcal{A}$ performing this measurement is statistically represented, see section 4.2, by

- probabilities for concrete results $p(\alpha_i|\rho)$;
- transformations of the initial state corresponding to the concrete results of measurement,

$$\rho \to \rho_{\alpha_i}. \tag{4.66}$$

The rigorous mathematical description of this statistical output leads to the notion of *a quantum instrument*, see section 4.12.1.

Mixing law. In the quantum operational formalism it is assumed that these probabilities, $p(\alpha_i|\rho)$, satisfy the *mixing law*. We remark that, for any pair of states (density operators) ρ_1, ρ_2 and any pair of probability weights $q_1, q_2 \geq 0, q_1 + q_2 = 1$, the convex combination $\rho = q_1\rho_1 + q_2\rho_2$ is again a state (density operator). In accordance with the mixing law any apparatus produces probabilities such that

$$p(\alpha_i|q_1\rho_1 + q_2\rho_2) = q_1 p(\alpha_i|\rho_1) + q_2 p(\alpha_i|\rho_2). \tag{4.67}$$

This is a very natural probabilistic assumption.

Composition of the apparatuses.

It is natural to assume that after measurement of an observable a experimenter can perform measurement of another observable b. In general the result of b depends on the result of preceding measurement of a.

Such a sequence of measurements is represented as a new apparatus, the composition of the apparatuses $\mathcal{A}$ and $\mathcal{B}$: $\mathcal{BA}$. Its outputs are ordered pairs of results (α_i, β_j). It is postulated that the corresponding output probabilities and states are determined as

$$p((\alpha_i, \beta_j)|\rho) = p(\beta_j|\rho_{\alpha_i})p(\alpha_i|\rho); \qquad (4.68)$$

$$\rho_{(\alpha_i, \beta_j)} = (\rho_{\alpha_i})_{\beta_j}. \qquad (4.69)$$

The law (4.68) can be considered as the quantum generalization of the Bayes rule, but for states and not simply probabilities. The law (4.69) is the natural composition law.

4.12.1 *Davis-Levis instruments*

Consider a quantum apparatus $\mathcal{A}$. Its statistical output determines the following instrument:

$$\mathcal{E}(\alpha_i)\rho = p(\alpha_i|\rho)\rho_{\alpha_i} \qquad (4.70)$$

and, for a subset Γ of O, where $O = \{\alpha_1, ..., \alpha_m\}$ is the set of all possible results of measurement, we set

$$\mathcal{E}(\Gamma)\rho = \sum_{\alpha_i \in \Gamma} \mathcal{E}(\alpha_i)\rho = \sum_{\alpha_i \in \Gamma} p(\alpha_i|\rho)\rho_{\alpha_i}. \qquad (4.71)$$

We point to the basic feature of this map:

$$\mathrm{Tr}\mathcal{E}(O)\rho = \sum_{\alpha_i \in O} p(\alpha_i|\rho)\mathrm{Tr}\rho_{\alpha_i} = 1. \qquad (4.72)$$

For each concrete result α_i, $\mathcal{E}(\alpha_i)$ maps density operators to linear operators (in the infinite dimensional case, these are trace-class operators, but we proceed in the finite dimensional case, where all operators have finite traces).

The mixing law implies that, for any $\Gamma \subset O$,

$$\mathcal{E}(\Gamma)(q_1\rho_1 + q_2\rho_2) = q_1\mathcal{E}(\Gamma)\rho_1 + q_2\mathcal{E}(\Gamma)\rho_2. \qquad (4.73)$$

As was shown by Ozawa [212], under the assumption on the existence of composition of the apparatuses any such a map $\mathcal{E}(\Gamma) : D(H) \to L(H)$ can be extended to a linear map (superoperator)

$$\mathcal{E}(\Gamma) : L(H) \to L(H) \qquad (4.74)$$

such that:

- each $\mathcal{E}(\Gamma)$ is positive, i.e., it transfers the set of positive-semidefinite operators into itself;
- $\mathcal{E}(O) = \sum_i \mathcal{E}(\alpha_i)$ is trace preserving:

$$\mathrm{Tr}\mathcal{E}(O)\rho = \mathrm{Tr}\rho. \tag{4.75}$$

The latter property is a consequence of (4.72).

Thus, the two very natural and simple assumptions, the mixing law for probabilities and the existence of composite apparatuses, have the fundamental mathematical consequence, the representation of the evolution of the state by a superoperator (4.74).

In quantum physics such maps are known as *Davis-Levis* (DL) [66] quantum operations. This notion of the quantum operation is more general than the notion used nowadays. The latter is based on complete positivity, instead of simply positivity, see section 4.12.2 for the corresponding definition and discussion.

Thus, each measurement with the result α_i induces the back-reaction which can be formally represented as a DL quantum operation. In these terms

$$\rho_{\alpha_i} = \frac{\mathcal{E}(\alpha_i)\rho}{\mathrm{Tr}\mathcal{E}(\alpha_i)\rho}. \tag{4.76}$$

We remark that the map $\Gamma \to L(L(H))$, from subsets of the set of possible results O into the space of superoperators, is additive:

$$\mathcal{E}(\Gamma_1 \cup \Gamma_2) = \mathcal{E}(\Gamma_1) + \mathcal{E}(\Gamma_2), \quad \Gamma_1 \cap \Gamma_2 = \emptyset. \tag{4.77}$$

This is a measure with values in the space $L(L(H))$. Such measures are called (DL) instruments [66].

The probabilities for results of measurements can be represented with the aid of POVMs. Take an instrument $\mathcal{E}$, where, for each $\alpha_i \in O, \mathcal{E}(\alpha_i) :$ $L(H) \to L(H)$ is a superoperator. Then we can define the adjoint operator $\mathcal{E}^\star(\alpha_i) : L(H) \to L(H)$. Set

$$M_{\alpha_i} = \mathcal{E}^\star(\alpha_i)I,$$

where $I : H \to H$ is the unit operator. Then, since

$$p_{\alpha_i} = \mathrm{Tr}\mathcal{E}(\alpha_i)\rho = \mathrm{Tr}\, I; \mathcal{E}(\alpha_i)\rho = \langle I|\mathcal{E}(\alpha_i)\rho\rangle$$

$$= \langle \mathcal{E}^\star(\alpha_i)I|\rho\rangle = \mathrm{Tr}(\mathcal{E}^\star(\alpha_i)I)\rho = \mathrm{Tr}M_{\alpha_i}\rho.$$

By using the properties of an instrument it is easy to show that M_{α_i} is POVM. Thus, for each apparatus $\mathcal{A}$, its probability output can be represented by POVM.

4.12.2   *Complete positivity*

Nowadays theory of DL-instruments is considered old-fashioned; the class of such instruments is considered to be too general: it contains mathematical artifacts which have no relation to real physical measurements and state transformations as back-reactions to these measurements. Modern theory of instruments is based on the extendability postulate, e.g., [211], [212]:

For any apparatus $\mathcal{A}_S$, for measurement of an observable a on a system S and any system $\tilde{S}$ noninteracting with S there exists an apparatus $\mathcal{A}_{S+\tilde{S}}$ for measurement on the compound system $S + \tilde{S}$ such that

- $p(\alpha_i | \rho \otimes r) = p(\alpha_i | \rho)$;
- $(\rho \otimes r)_{\alpha_i} = \rho_{\alpha_i} \otimes r$

for any state ρ of S and any state r of $\tilde{S}$.

This postulate is very natural: if, besides the quantum system S which is the object of measurement, there is (somewhere in the universe) another system $\tilde{S}$ which is not entangled with S, i.e., their joint pre-measurement state has the form $\rho \otimes r$, then the measurement on S with the result α_i can be considered as measurement on $S + \tilde{S}$ as well with the same result α_i. It is clear that the back-reaction cannot change the state of $\tilde{S}$. Surprisingly this very trivial assumption has tremendous mathematical implications.

Since we proceed only in the finite dimensional case, the corresponding mathematical considerations are simplified. Consider an instrument $\mathcal{E}_S$ representing the state update as the result of the back-reaction from measurement on S. For each Γ, this is a linear map from $L(H) \to L(H)$, where H is the state space of S. Let W be the state space of the system $\tilde{S}$. Then the state space of the compound system $S + \tilde{S}$ is given by the tensor product $H \otimes W$. We remark that the space of linear operators in this state space can be represented as $L(H \otimes W) = L(H) \otimes L(W)$. Then the superoperator $\mathcal{E}_S(\Gamma) : L(H) \to L(H)$ can be trivially extended to the superoperator $\mathcal{E}_S(\Gamma) \otimes I : L(H \otimes W) \to L(H \otimes W)$. It is easy to prove that the quantum instrument representing the apparatus for measurements on $S + \tilde{S}$ has to have this form $\mathcal{E}_{S+\tilde{S}}(\alpha_i) = \mathcal{E}_S(\alpha_i) \otimes I$. Hence, this operator also has to be positive. We remark that if the state space W has the dimension k, then the space of linear operators $L(W)$ can be represented as the space of $k \times k$ matrices which is further denoted as $\mathbf{C}^{k \times k}$.

Formally, a superoperator $T : L(H) \to L(H)$ is called *completely positive* if it is positive and each of its trivial extensions $T \otimes I : L(H) \otimes \mathbf{C}^{k \times k} \to$

$L(H) \otimes \mathbf{C}^{k \times k}$ is also positive. There are natural examples of positive maps which are not completely positive.

A CP quantum operation is a DL quantum operation which is additionally completely positive; a CP instrument is based on CP quantum operations representing back-reactions to measurement. As was pointed out, in modern literature only CP quantum instruments are used, so they are simply called quantum instruments.

The main mathematical feature of CP quantum operations is that the class of such operations can be described in a simple way, namely, with the aid of the *Kraus representation* [212], [48]:

$$T\rho = \sum_j V_j \rho V_j^\star, \tag{4.78}$$

where (V_j) are some operators acting in H. Hence, for a CP instrument, we have: for each $\alpha_i \in O$, there exist operators $(V_{\alpha_i j})$ such that

$$\mathcal{E}(\alpha_i)\rho = \sum_j V_{\alpha_i j} \rho V_{\alpha_i j}^\star. \tag{4.79}$$

Thus

$$\rho_{\alpha_i} = \frac{\sum_j V_{\alpha_i j} \rho V_{\alpha_i j}^\star}{\mathrm{Tr} \sum_j V_{\alpha_i j} \rho V_{\alpha_i j}^\star}, \tag{4.80}$$

where the trace one condition (4.72) implies that

$$\sum_i \sum_j V_{\alpha_i j} V_{\alpha_i j}^\star = I. \tag{4.81}$$

The corresponding POVMs M_{α_i} can be represented as

$$M_{\alpha_i} = \sum_j V_{\alpha_i j}^\star V_{\alpha_i j}. \tag{4.82}$$

This is a really elegant mathematical representation. However, it might be that this mathematical elegance and not the real physical situation have contributed to widespread of CP in quantum information theory.

Chapter 5

Quantum and Contextual Probability

We begin with the following terminological remark. In this chapter, the term "classical probability" is used for probability represented mathematically by the measure-theoretic model of Kolmogorov (so to say, Kolmogorovean probability); the term "quantum probability" is used for theory of probability based on the complex Hilbert representation and Born's rule connecting states with probabilities. We recall that in Chapter 1 we also considered another classical probability model developed by von Mises. In consideration of the present chapter we shall proceed solely with the Kolmogorov probability model, see [133], [156] for the corresponding presentation based on the von Mises probability model.

Starting with analysis of the probabilistic structure of the two-slit experiment, we shall enlighten the role played by *experimental contexts* in QM. Here we follow N. Bohr[1], but his viewpoint on the role played by experimental arrangement will get the explicit contextual probabilistic representation.

Surprisingly the first explicit and logically structured discussion representing the *contextual probabilistic viewpoint on the two-slit experiment* and emphasizing its nonclassical probabilistic structure was presented by R. Feynman [88]. This is really surprising, because Feynman was pure physicist with the minimal background in probability theory. Nevertheless, in [88] he discussed foundations of classical and quantum probabilities and their incompatibility.

At the same time this is not surprising by taking into account brightness of Feynman's mind. He demonstrated many times his ability to heuristic

---

[1] "Strictly speaking, the mathematical formalism of quantum mechanics and electrodynamics merely offers rules of calculation for the deduction of expectations pertaining to observations obtained under well-defined experimental conditions specified by classical physical concepts", [39].

vision of the novel and yet formally unrepresentable physical phenomena. We can point to the path integral representation of QM. The Feynman representation of the complex probability amplitudes in the form of integrals over the space of all possible trajectories of a quantum particle definitely deviated from the Copenhagen interpretation. By the latter the space-time causal representation of quantum phenomena is strictly forbidden, see sections 6.3, 6.4.[2] We also recall that R. Feynman was one of the pioneers of the quantum computation project. It is interesting that again the main foundational motivation of impossibility to solve problems of QM on classical computer in reasonable (polynomial) time was of the probabilistic nature. In fact, he again appealed to the nonclassical structure of quantum probability which can be formally expressed with the aid of "negative probabilities". Such "probabilities" cannot be represented in classical probability theory, but the wave function can be treated as a kind of encryption of them. (In this book we do not plan to discuss the issue of "negative probability" in QM, see section 3.5. We did this in very detail in the monographs [133], [156].)

Coming back to the two-slit experiment, we remark that Feynman even did not know about the Kolmogorov measure-theoretic model, he used the Laplacian probability theory. And, before Feynman, no expert in probability discovered "violation of laws of classical probability theory" in quantum experiments. The situation did not change so much even after the publication of the book of Feynman and Hibbs [88]. It seems that the first mathematically rigorous formulation of Feynman's consideration was presented

---

[2]When R. Feynman presented the first time his path integral approach to QM at the seminar of N. Bohr, he was abusively criticized. Bohr emphasized that the path integral representation of quantum amplitudes is unphysical. Roughly speaking it was strongly recommended to Feynman to disappear from the quantum landscape with his theory. Others recommended Feynman to establish a closer contact with W. Pauli who was more open-minded and less dogmatic than Bohr; Feynman also got to know that the best way to start is to invite Pauli to joint drink. Feynman bought a bottle of very good cognac and at some occasion invited Pauli to drink it. These two great men immediately appreciated open-mindedness of each other. And Pauli who was very good to find unexpected solutions, told Feynman: "just call them virtual trajectories." Then Pauli also "explained" to Bohr that in his talk Feynman misinterpreted his formalism. The second talk of Feynman at Bohr's seminar was the real triumph. Nowadays the path intergral approach is widely used in QM, string theory, cosmology. Typically physicists do not take care about using the right interpretation of such integrals. (They are neither take care about difficulties in defining such "integrals" in rigorous mathematical framework, cf. [123], [124], [236], [132].) This is one of a variety of the "quantum psychological effects". Discussing the foundational issues one typically emphasizes that he uses the Copenhagen interpretation of QM, but by doing concrete calculations he rarely remember about this.

in the series of my works [141], [142], [161], [152], [153] and summarized in the monograph [133] (contextual probabilistic treatments of other quantum experiments can be found in papers [135], [137], [140]). In section 5.1 we briefly repeat Feynman's argument that the quantum formalism is incompatible with laws of classical probability. We modify the original Feynman consideration based on violation of the law of additivity of probability. We proceed with the *formula of total probability* (FTP), section 1.6. FTP plays the important role in probability inference. Therefore its violation reflects better novel features of the quantum probability inference, cf. sections 6.7.3, 6.9 (QBism and Växjö interpretation: the quantum formalism as the special machinery of nonclassical probability inference).

In section 5.2 we compare the classical wave interference with *"interference of probabilities"* in QM and point to similarities and dissimilarities. We also discuss possibilities of treating QM as emergent from some subquantum theory of the classical field type by presenting the views of Schrödinger, Einstein and Infeld, and the author of this book. By representing QM as the special image of classical field theory (and such a possibility exists - section 5.2) we, of course, reduce quantum "interference of probabilities" to classical interference of physical waves propagating in space, modeled as $\mathbf{R}^3$. However, I do not claim that the physical wave interpretation of quantum interference is the right interpretation of this phenomenon. In section 5.4 we show that *"complex waves of probability"* can appear as a special mathematical representation of sufficiently general contextual probabilistic data. Thus it seems that such "waves of probability" need not be reduced to physical waves. The formalism of quantum "waves of probability" might be applicable to statistical phenomena outside of physics, e.g., for cognition, psychology, sociology, economics, politics, Chapter 9. Here definitely the model of interference of probabilities based on waves propagating in physical space is improper. However, "waves of probability" are very useful mathematical tool. Unfortunately, the problem of the wave representation of statistical data of any origin is mathematically very difficult. I was able to solve it only for dichotomous observables [161], section 5.2.

Before proceeding to representation of contextual statistical data in complex Hilbert space (see section 5.4), in section 5.3 we generalize Feynman considerations for the two-slit experiment to the abstract multi-context experimental situation and derive generalized FTP, the additive perturbation of classical FTP. The additive perturbation can be treated as an abstract probabilistic analog of the *interference term*.

The most general contextual probabilistic model will be presented in section 5.5. Here we shall also analyze inter-relation of the contextual and classical (Kolmogorov) probabilistic models. The use of the Kolmogorov probability space to represent multi-contextual situations implies the possibility to define the *joint probability distributions of observables* which are defined for different contexts. We sharply distinguish *observational and probabilistic (in)compatibilities*.

The first one was analyzed by Bohr and Heisenberg and formalized in the form of the principle of complementarity. In our contextual model, contexts $C_1, ..., C_m$, are observationally compatible if there exists context, say C, permitting the joint measurement of observables related to these contexts. The probabilistic compatibility, the existence of the joint probability distribution of observables, is closely related to observational compatibility. However, the existence of the probability distribution might be just a mathematical artifact which does not imply compatibility in the sense of Bohr-Heisenberg.

It is clear from the experimental situations of the two-slit experiment and experiments on violation of the Bell-type inequality that there exist experimental contexts which are not only observationally, but even probabilistically incompatible. This motivates us to use the terminology "non-Kolmogorovean probability" for contextual probabilistic models containing probabilistically incompatible contexts. Following Feynman [88], one can speak about violation of the laws of classical probability theory. This viewpoint was especially clearly presented and emphasized in works and numerous talks of L. Accardi [2] - [4] who invented the terminology *non-Kolmogorovean probability models*. This terminology is convenient as it enlightens impossibility of the straightforward embedding of some data into a single Kolmogorov probability space. However, the interrelation of contextual and Kolmogorovean probabilities is more complicated than it was thought by Feynman and Accardi and myself, see [170]. Thus the terminology "non-Kolmorovean probability" has to be used with caution, see section 5.3.3.

5.1   Probabilistic Structure of Two-Slit Experiment

The two-slit experiment is the basic example demonstrating that QM describes statistical properties in microscopic phenomena, to which the classical probability theory *seems to be not applicable*, see, e.g., Feynman et al. [88]. In this section, we consider the experiment with the symmetric setting: the source of photons is located symmetrically with respect to two slits, Fig. 5.1.

Consider the following pair of observables a and b. We select a as the "slit passing observable," i.e., $a = 0, 1$, see Fig. 5.1, (we use indices $0, 1$ to be close to qubit notation) and observable b as the position on the photo-sensitive plate, see Fig. 5.1. We remark that the b-observable has the continuous range of values, the position x on the photo-sensitive plate. We denote $P(a = i)$ by $P(i)$ ($i = 0, 1$), and $P(b = x)$ by $P(x)$. Physically the a-observable corresponds to measurement of position (coarse grained to "which slit?") and the b-observable represents measurement of momentum.

In quantum foundational studies, various versions of the two-slit experiment have been successfully performed, not only with photons, but also with electrons and even with macroscopic molecules [259]. All those experiment demonstrated matching with predictions of QM. Experimenters reproduce the interference patterns predicted by QM and calculated by using the wave functions.

The probability that a photon is detected at position x on the photosensitive plate is represented as

$$P(x) = \left| \frac{1}{\sqrt{2}} \psi_0(x) + \frac{1}{\sqrt{2}} \psi_1(x) \right|^2$$
$$= \frac{1}{2} |\psi_0(x)|^2 + \frac{1}{2} |\psi_1(x)|^2 + |\psi_0(x)| |\psi_1(x)| \cos \theta, \qquad (5.1)$$

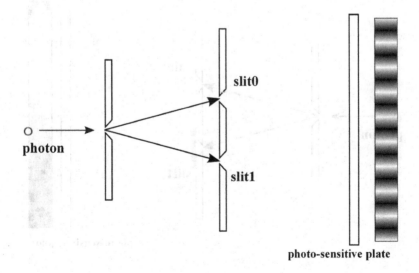

Fig. 5.1   Context with both slits are open.

where ψ_0 and ψ_1 are two wave functions, whose squared absolute values $|\psi_i(x)|^2$ give the distributions of photons passing through the slit $i = 0, 1$, see Figs. 5.2 and 5.3. Here we explored the rule of addition of complex probability amplitudes, a quantum analog of the rule of addition of probabilities. This rule is the direct consequence of the linear space structure of quantum state spaces.

The term

$$|\psi_0(x)| \, |\psi_1(x)| \cos\theta$$

implies the interference effect of two wave functions. Let us denote $|\psi_i(x)|^2$ by $P(x|i)$, then Eq. (5.1) is represented as

$$P(x) = P(0)P(x|0) + P(1)P(x|1) + 2\sqrt{P(0)P(x|0)P(1)P(x|1)} \cos\theta. \quad (5.2)$$

Here the values of probabilities $P(0)$ and $P(1)$ are equal to $1/2$ since we consider the symmetric settings. For general experimental settings, $P(0)$ and $P(1)$ can be taken as the arbitrary nonnegative values satisfying $P(0) + P(1) = 1$. In the above form, the classical probability law (FTP)

$$P(x) = P(0)P(x|0) + P(1)P(x|1), \quad (5.3)$$

is violated, and the term of interference $2\sqrt{P(x|0)P(0)P(x|1)P(1)} \cos\theta$ specifies the violation.

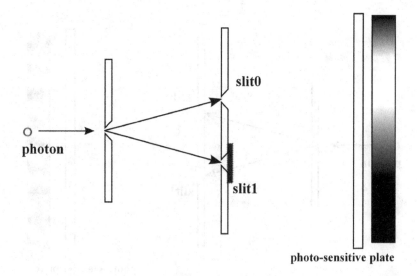

Fig. 5.2   Context with one slit closed (I).

Typically this violation of FTP is presented as one of quantum mysteries and "the violation of laws of classical probability" is coupled with the exotic features of quantum particles. We do not share this view. In our opinion, the violation of FTP is a consequence of the special contextual structure of the two-slit experiment (in fact, a group of experiments), see section 5.3.

5.2 Quantum versus Classical Interference

Here we discuss the difference between the use of the notion of interference in classical optics and QM. Classically this is interference of two physical waves (classical electromagnetic fields) propagating from two open slits. (The two-slit experiment, for classical light and photons, will be used as an illustration.) In QM this is interference of two possibilities, corresponding to (virtual) passing through two slits.

In classical optics one need not to compare "outputs of possibilities". It is enough to perform the experiment just for one context: both slits are open. Here one can see the interference pattern, the result of addition of two physical waves, and this is the final result.

In QM, to compare possibilities (their statistical outputs), the experiment has to be multi-contextual, $C_i, i = 0, 1$, only the ith slit is open,

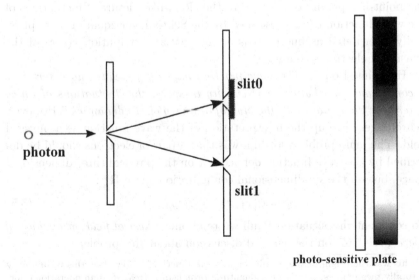

Fig. 5.3   Context with one slit closed (II).

and C_{01}, both slits are open, see Figs. 5.2, 5.3, and 5.1. Comparison of possibilities is represented as comparison of the corresponding probability distributions $P(x|i), P(x)$. In the contextual notations they can be written as

$$p^b_{C_i}(x) \equiv P(b = x|C_i), p^b_{C_{01}}(x) \equiv P(b = x|C_{01}).$$

Here conditioning is not the classical probabilistic event conditioning, but context conditioning, see sections 5.3 and 5.5.

Before continuing with analysis of multi-contextual experiments, section 5.3, we analyze how far one proceeds in the interference experiments by using the analogy with interference of classical waves.

5.2.1 Quantum waves?

We remark that the wave function was invented by Schrödinger who treated it as a physical wave.[3] In his paper *An undulatory theory of the mechanics of atoms and molecules* [230] he treated the wave function as a physical field, the amplitude of the density of electron's charge, i.e.,

$$r(t, x) = e|\psi(t, x)|^2 \tag{5.4}$$

and the wave function (as in modern QM) was normalized by 1, $\int |\psi(t, x)|^2 dx = 1$. Thus, for Schrödinger, QM was simply transition from the point-like picture of electron to the cloud-like picture. Fluctuations of the wave function $\psi(t.x)$ described by the Schrödinger equation were physically interpreted as fluctuations of the spatial distribution $r(t, x)$ of the density of electron's charge.

He pointed out: *"The wave-function physically means and determines a continuous distribution of electricity in space, the fluctuations of which determine the radiation by the laws of ordinary electrodynamics."* However, Schrödinger gave up the interpretation of the wave function as a physical field. The main problem for him was that a pair of electrons should be described by the wave function defined not on the physical three dimensional space, but on the six-dimensional configuration space $\mathbf{R}^6$:

$$\psi = \psi(x_1, x_2, x_3, y_1, y_2, y_3). \tag{5.5}$$

To comment this situation Pauli wrote about *"physical field on unphysical space"*, see [167] on the detailed discussion about this problem.

---

[3]This was the origin of the terminology "wave function". Nowadays this terminology is totally meaningless, because the quantum formalism is treated as an operational formalism to count detection probabilities. In accordance with the modern viewpoint, QM is not about waves.

We remark that nowadays such multi-dimensional spaces are used on the equal grounds with the three dimensional physical space, for example, in *string theory*, where, e.g., $D = 26$ is treated as a physical dimension. Moreover, theory of superstrings is based on multi-dimensional superspaces, where coordinates are not real numbers, but belong to supercommutative superalgebras [132]. In string theory, cosmology, and theory of complex disordered systems (spin glasses) even more exotic configuration spaces, namely, based on the fields of p-adic numbers $\mathbf{Q}_p$, where $p = 2, 3, ..., 1997, 1999, ...$, are prime numbers, are widely applied [127], [246], [126]. (It is not clear what configuration space is more exotic $\mathbf{Q}_p^3$ or $\mathbf{R}^6$?) I guess that if such configuration spaces were widely used at the beginning of 1920s, it might be that Schrödinger would not be so strongly disappointed by appearance of "physical fields on unphysical spaces" and would not give up the physical field interpretation of the quantum wave function so easily.

In any event until his last days he dreamed for total exclusion of particles from quantum theory and creation of a classical field model covering the quantum phenomena. The same viewpoint was presented in very details in the book of Einstein and Infeld [86]. Einstein until his last days worked on nonlinear classical field model unifying QM and general relativity. However, his long term and tremendous efforts were not successful. It seems that he did not publish anything about his last studies. (My information is based on the private communication of a relative of Einstein who worked with his archive.)

5.2.2 *Prequantum classical statistical field theory*

This section is devoted to essentials of one special model of the field type explaining quantum interference (and other quantum statistical patterns) in the classical manner, as the result of addition of physical waves. This material is more special and this section deviates from the main topic of this book. Therefore the reader may jump to section 5.3.

Stimulated by the book of Einstein and Infeld [86] and the initial ideas of Schrödinger, I created [167] my own subquantum model which is known as *prequantum classical statistical field theory* (PCSFT). Here QM appears as the operational representation of correlations of classical random fields. We point out that the PCSFT-interpretation of the wave function differs from the original Schrödinger interpretation. The wave function by itself is not a physical field. *It represents the covariance operator of classical (pre-*

quantum) random field. This covariance operator B is a singular integral operator and its kernel can be expressed in terms of the wave function.

To illustrate the possibility to handle quantum entanglement in the PCSFT framework, consider the case of a bipartite quantum system $S = (S_1, S_2)$ with the wave function $\psi(x, y), x, y \in \mathbf{R}^3$. In PCSFT quantum systems are operational entities representing various prequantum random fields. Thus in PCSFT by saying about a bipartite quantum system, we, in fact, consider a pair of (in general correlated) random fields:

$$\phi(x, y) = (\phi_1(x), \phi_2(y)). \tag{5.6}$$

The fields $\phi_1(x)$ and $\phi_2(y)$ represent the quantum systems S_1 and S_2, respectively. To match with notations of classical probability theory, we have to use the chance parameter ω and write the field and its components as random variables: $\phi(x, y; \omega) = (\phi_1(x; \omega), \phi_2(y; \omega))$.

The covariance operator B of the pair of prequantum classical random fields (5.6) can be represented in the matrix form:

$$B = \begin{pmatrix} B_{11} & B_{12} \\ B_{21} & B_{22} \end{pmatrix}. \tag{5.7}$$

Here the operator B_{ii} represents self-correlations in the field $\phi_i, i = 1, 2$, and the operator B_{12} represents correlations between the fields ϕ_1 and ϕ_2. We remark that $B_{21} = B_{12}^\star$ (the adjoint operator).

The block B_{12} of the covariance operator B is the integral operator with the kernel coinciding with the wave function ψ :

$$B_{12}(x, y) = \psi(x, y), \text{ i.e., } B_{12}f(x) = \int \psi(x, y)f(y)dy. \tag{5.8}$$

This is the key-formula of the PCSFT-representation for bipartite systems. Thus we solved Schrödinger's problem of the interpretation of the wave function $\psi(x, y)$ of a bipartite system as a classical physical field on "unphysical space" $\mathbf{R}^6$. The inter-correlation function of a pair of two random physical fields, see (5.6), is really defined on $\mathbf{R}^6$.

To determine the diagonal blocks of the covariance operator B, we recall that in QM, the state of a bipartite system $S = (S_1, S_2)$ determines the states of its subsystems $S_i, i = 1, 2$ (but not vice versa) with the aid of the operation of the partial trace. For a pure state represented by the L^2-function $\psi(x, y)$, the states of S_i are always mixed states, ρ_i. They are integral operators with the kernels:

$$\rho_1(x_1, x_2) = \int \psi(x_1, y)\overline{\psi(x_2, y)}dy, \tag{5.9}$$

$$\rho_2(y_1, y_2) = \int \psi(x, y_1)\overline{\psi(x, y_2)}dx. \tag{5.10}$$

The first guess is that the diagonal blocks of the covariance operator B of the classical random field $\phi(x, y)$ coincide with the density operators of the subsystems S_i

$$B_{ii} = \rho_i, , \text{ i.e., } B_{11}(x_1, x_2) = \rho_1(x_1, x_2), B_{22}(y_1, y_2) = \rho_2(y_1, y_2). \tag{5.11}$$

This choice leads to matching of quantum and classical random field averages as well as correlations for the subsystems S_i. However, life is not so easy and we can proceed with (5.11) only for factorizable ψ, i.e., $\psi(x, y) = \psi_1(x)\psi_2(y)$. For an entangled state ψ, the operator B with the blocks determined by (5.8) and (5.11) *is not positive-semidefinite*.

To solve this problem, we have to take into account that in PCSFT the *background field* (so to say "zero point field") plays the crucial role. This background component of a random field is not represented by the wave function. However, it also has to be taken into account in the covariance operator of a prequantum random field (to obtain a mathematically consistent model). By ignoring the background field we were obtain "covariance operators" which are not positive-semidefinite. By taking into account the contribution of the background field of the white noise type, we define the diagonal blocks of B as integral operators with the kernels

$$B_{11}(x_1, x_2) = \rho_1(x_1, x_2) + \epsilon\delta(x_1 - x_2), \tag{5.12}$$

$$B_{22}(y_1, y_2) = \rho_1(y_1, y_2) + \epsilon\delta(y_1 - y_2), \tag{5.13}$$

where $\epsilon > 0$ represents the strength of the background field. It can be proved that, for a sufficiently strong background field, the integral operator B with the diagonal blocks given by (5.12), (5.13) and the off-diagonal block given by (5.8) is positive-semidefinite. We remark that, in contrast to the operator with the blocks (5.11), (5.8), this operator is not of the Hilbert-Schmidt type; it is a singular integral operator. The corresponding, e.g., Gaussian, random field (5.6) is singular, it takes values, not in the L^2-space, but in a proper space of distributions on $\mathbf{R}^3$, e.g., in the space of Schwartz distributions. Thus my mathematical model says that "prequantum random fields", i.e., fields which we label as quantum systems, are extremely singular.

We remark that we considered the background contribution of the white noise type just for simplicity. Our mathematical model of PCSFT is indifferent with respect to the choice of a background component. A more complex random fields can be used as well. The only restriction on the random background is that its presence should transform the "pure ψ-component of the covariance operator" given by (5.8), (5.11) into a positive-semidefinite operator.

We also remark that in general the covariance operator does not determine a random field. We point out that in PCSFT prequantum random fields have zero average. However, even by assuming that average is zero and the covariance operator is fixed and given by (5.12), (5.13), (5.8) we do not fix the random field (5.6), where $\phi = \phi(x, y; \omega)$ and ω is the chance variable. Thus a variety of random fields can reproduce the same quantum correlation. To fix the random field, we can assume that it is Gaussian. And there are some physical reasons supporting this assumption [167], i.e., that "quantum systems" are symbolic representations of Guassian prequantum fields. Another possibility is that the same wave function represents a variety of experimental contexts with very different random fields behind "quantum systems". For a moment, we cannot solve this problem. However, if PCSFT rightly represents physical reality, then in future it would be possible to measure prequantum random fields. We would be able to determine their type, e.g., to test whether they are Gaussian or not.

We emphasize that PCSFT represents all quantum correlations as correlations of classical random fields [167] – in spite of all no-go theorems, including the Bell inequality theorem. PCSFT "beats the latter" by unusual reason: *in PCSFT there are no systems.* Hence, there are no hidden variables which can be assigned to systems. This model can be considered as a realist subquantum model which is not of the hidden variables type, see sections 6.1.1 and 6.1.2.[4]

The quantum formalism arises as an approximation of the classical field formalism in the following natural way. A nonlinear functional $F(\phi)$ of a classical field ϕ is expanded by using the Taylor formula at the point

---

[4] One might say that PCSFT "beats Bell's no-go" as the result of the presence of the background field. However, this would not be totally correct. The mathematical model is more complicated than it was briefly presented above. In fact, we can speak about background field and pulses emitted by a source only heuristically. In the rigorous mathematical model we cannot separate a pulse from the background field, on the level of fields. We can only separate their contributions on the level of covariances, as two terms in the covariance operator B given by (5.12), (5.13), (5.8), see [167] for the rigorous mathematical details. Thus we can say "electrons, photons, ..., neutrons do not exist separated from the background field". In PCSFT the "system ideology" behind conventional QM and Bell's type hidden variable reasoning is not applicable. Therefore we can say that with creation of PCSFT the dream of Einstein and Infeld [86] and Schrödinger [228], [230] for quantum theory as a pure field theory, i.e., without particle-like systems, came true.

$\phi = 0$, up to the second derivative. Then a prequantum random field is used as variable, i.e., we consider a random functional and its Taylor expansion. Finally, this expansion is averaged. The contribution of the first order term is vanished under the assumption that prequantum random fields have zero average. The contribution of the second order gives the quantum average. The second derivative $F''(0)$, which is always a Hermitian operator, is operationally treated as a quantum observable $A = F''(0)$. The main mathematical complication is that the state space of classical fields is infinite-dimensional. Hence, we have to use the differential calculus, in particular, the Taylor formula on infinite dimensional space. However, the corresponding mathematical methods are well-developed, e.g., to serve optimization theory. And I successfully applied them to emerge QM from classical field theory [167].

The next step in development of PCSFT was creation of the corresponding measurement theory. The physical structure of the detection model is very simple. A continuous random field is transferred into discrete clicks of detectors with the aid of threshold passing mechanism. When the power of the incoming random signal at the point of detection becomes higher than detector's threshold, the detector produces a click. Of course, a detector has to be of the threshold type. And the latter seems to be the case: all quantum single photon detectors are of the threshold type. Moreover, thresholds are also used in measurements for massive systems, e.g., electrons and neutrons. It seems that it is impossible to proceed without detection thresholds.

However, conventionally thresholds are treated not as the basic elements of detection, transferring continuous fields into discrete clicks, but as noise-cutting elements. Quantum signals are very noisy and only by putting thresholds can one cut the main part of noise.

In PCSFT, the above simple and natural picture of threshold detection of continuous signals is shadowed by necessity to take into account the contribution of the background field. Thus the event that a detector clicks need not be a consequence of a high power pulse emitted from a source of signals (which is operationally presented as a quantum state). A click is a consequence of the joint contribution of this pulse and the random background field. And the latter can not only exhaust pulse's impact on a detector, but also have a destructive interference contribution. All fields are mathematically described in linear space. Fields' energies are combined not additively, but similarly to QM via addition of fields' components. However, as was already pointed out, the mathematical situation is more complicated:

a pulse does not exist separately from the background field.

This similarity, the linear space representation of classical fields and quantum states will always disturb those who do not take physics formally. *Can quantum interference of probability amplitudes be emergent from classical wave interference?* It seems so natural to answer "yes" that from year to year (already practically one hundred years) people try to obtain QM as an operational theory emergent from some sort of classical field theory.

This simple mechanism, discretization of continuous signals with the aid of the threshold type detectors, works well to reproduce the basic predictions of QM from PCSFT, including violation of the Bell inequality. The main difficulty of PCSFT is that, although it represents easily bipartite correlations, the classical field model for tripartite correlations becomes very complicated [167]. Here the main problem is even not mathematical complexity of calculations, but a very technical and heuristically unpleasant structure of extraction of the terms of classical tripartite correlations corresponding to quantum tripartite correlations. And already the case of quadripartite correlations is so complicated mathematically that I was not able to complete calculations. There can be mentioned a few possible explanations of this situation:

(1) The complexity of multi-signal correlations corresponds to the real physical situation. QM presents a very simple approximate picture, but the real physical picture given by classical fields is really extremely complex. Thus one has to continue to solve computational and interpretational difficulties.

(2) Generalization from heuristically natural classical field model for quantum bipartite correlations to tripartite correlations proposed in [167] does not match the real physical situation. New and simpler mathematically generalization can be found.

(3) In general PCSFT is a misleading approach to emerge QM from a causal classical model. In the case of bipartite correlations it works successfully by chance.

5.3 Formula of Total Probability with Interference Term

5.3.1 *Context-conditioning*

In classical probability theory we assume that, for any pair of events A and B, it is possible to form the event $B \cap A$, "both events A and B take place." Suppose that these events correspond to measurements of two (discrete) random variables a and b. That is, we can consider the two families of the events

$$E_\alpha^a = \{\omega \in \Omega : a(\omega) = \alpha\}, E_\beta^b = \{\omega \in \Omega : b(\omega) = \beta\}.$$

In classical probability theory the event corresponding to the joint measurement of these random variables is always well defined and it is conjunction of the events E_α^a and E_β^b given by their set-theoretic intersection, $E_{\alpha\beta}^{ab} = E_\alpha^a \cap E_\beta^b = \{\omega \in \Omega : a(\omega) = \alpha, b(\omega) = \beta\}$.

However, quantum observables are in general *incompatible*, i.e., it may be impossible to construct an experimental context for their joint measurement, then the event $E_{\alpha\beta}^{ab}$ is meaningless.

Therefore *conditioning with respect to an event* based on Bayes' formula (1.23) and exploring the operation of Boolean conjunction of events has a restricted domain of application. In the general situation conditioning has to be treated as conditioning with respect to *context of measurement* (section 5.5) and not event.

Hence, in the expression $P(B|C)$ the symbols B and C have to be treated in different ways. The first one, B, still denotes an event corresponding to the measurement for the value β of the random variable b. However, the symbol C is used to denote context for measurement of the b-observable. We now turn to FTP.

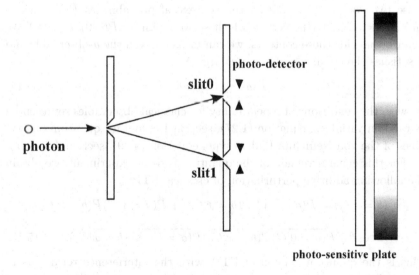

Fig. 5.4   Context with two detectors.

Can we generalize FTP by using context conditioning instead of Kolmogorovian event conditioning?

As was shown, see (5.2), in quantum physics the standard FTP can be violated. To simplify the introduction to the two-slit experiment, in section 5.1 we studied this experiment in the symmetric setting, the source was located symmetrically with respect to the slits. In such a setting we can use equal probabilities for passing through slits without further explanation.

To formalize the general situation, we introduce two (incompatible!) observables: b gives the point x of detection on the registration screen and $a = 0, 1$ gives information about which slit is passed. We recall that physically they represent the momentum and position observables (and the latter is discretized). Both observables are measured under the context $C \equiv C_{01}$, both slits are open, see Fig. 5.1 for b-measurement and Fig. 5.4 for a-measurement. The following contextual probabilities are obtained: $P(b = x|C)$ and $P(a = i|C)$, $i = 0, 1$. Besides the context C, two other contexts are involved: C_i, only i-th slit is open, $i = 0, 1$, Figs. 5.2 and 5.3, see also section 5.2. (In general figures have to be modified, the source need not be located symmetrically with respect to the slits. In section 5.2 we did not discuss and illustrate a-measurement (by Fig. 5.4), because for the symmetric location of the slits it is evident (a priori) that the probabilities $P(a = i|C) = 1/2$.) We also measure the b-observable with respect to these two contexts and obtain the contextual probabilities $P(b = x|C_0)$ and $P(b = x|C_1)$. (In section 5.1 these were simply $P(x|0), P(x|1)$.) We remark that, for these contexts, we can measure even the a-observable and it satisfies the condition of *repeatability*:

$$P(a = j|C_i) = \delta_{ij}, \tag{5.14}$$

cf. with the conditions of repeatability for classical observables represented as random variables, equation (1.27), section 1.5, and for quantum observables of the von Neumann-Lüders type, equation (4.29), section 4.3.

For this general version of the quantum two-slit experiment, we obtain the following additive perturbation of classical FTP:

$$P(b = x|C) = P(a = 0|C)P(b = x|C_0) + P(a = 1|C)P(b = x|C_1)$$

$$+ 2\cos\theta \sqrt{P(a = 0|C)P(b = x|C_0)P(a = 1|C)P(b = x|C_1)}. \tag{5.15}$$

This is the quantum version of FTP, with the interference term. As in section 5.1, we use the quantum formalism (the principle of superposition of wave functions and Born's rule).

As was already pointed in section 5.1, the main source of surprise by violation of classical FTP (and, hence, classical probability theory), see, e.g., Feynman et al. [88], is the application of event conditioning and the set-theoretical algebra for events. Motivated by (5.14) people (even experts in quantum foundations) typically try to identify contexts $C_i, i = 0, 1$, with the events $E_i^a = \{\omega \in \Omega : a(\omega) = i\}$ and context C with the event $E_0^a \cup E_1^a$. Under such identification, i.e., by assuming the possibility to use the measure-theoretic representation of probability, one also expects that classical FTP holds true:

$$P(b = x | E_0^a \cup E_1^a) = P(E_0^a)P(b = x|E_0^a) + P(E_1^a)P(b = x|E_1^a) \quad (5.16)$$

under the natural assumptions:

$$P(E_i^a) = P(E_i^a|E_0^a \cup E_1^a), \ P(b = x) = P(b = x|E_0^a \cup E_1^a).$$

Moreover, typically the set $E_0^a \cup E_1^a = \{\omega \in \Omega : a(\omega) = 0\} \cup \{\omega \in \Omega : a(\omega) = 1\}$ is identified with the complete set of elementary events Ω and (5.16) takes the simple form (see Chapter 1):

$$P(b = x) = P(E_0^a)P(b = x|E_0^a) + P(E_1^a)P(b = x|E_1^a). \quad (5.17)$$

5.3.2 Contextual analog of the two-slit experiment

Now we can repeat the previous contextual analysis in the abstract framework, i.e., without direct coupling to the two-slit experiment and QM. However, coming contextual probabilistic considerations are still simplified - to have a heuristically clear picture of the contextual description of measurements, see section 5.5 for the complete formal mathematical model.

There are two in general incompatible observables, say a and b. For our applications, it is sufficient to consider dichotomous observables, $a = \alpha_1, \alpha_2$ and $b = \beta_1, \beta_2$. (In the two-slit experiment the latter corresponds to coarse graining of the momentum observable.) There are given experimental context C and probabilities with respect to it, $P(a = \alpha_i|C)$ and $P(b = \beta_i|C)$.

There are also given two other contexts, so to say "$[a = \alpha_i]$-contexts", denoted by $C_{\alpha_i}, i = 1, 2$. In the classical probability model we are able to use the set-theoretic representation and represent C_α as $E_\alpha^a = \{\omega \in \Omega : a(\omega) = \alpha\}$. However, in the abstract contextual framework we have to determine these contexts by some probabilistic constraints. In the Kolmogorov model such probabilistic constraints are given by conditions of *repeatability*, equation (1.27), section 1.5; in the quantum model this is the condition of repeatability for observables of the von Neumann-Lüders type, equation (4.29), section 4.3. To determine "$[a = \alpha_i]$-contexts" C_α, we use their abstract contextual version:

$$P(a = j|C_{\alpha_i}) = \delta_{ij}, \tag{5.18}$$

cf. (5.14). (We do not assume the possibility of the measure-theoretic representation of these contexts. Thus the constraint (5.18), is not redundant to Boolean algebra.) Contexts $C_{\alpha_i}, i = 1, 2$, can be treated as *filtration contexts* corresponding to the outputs $a = \alpha_1, \alpha_2$ of measurement of the a-observable. In context C_{α_i} only output channel corresponding to the value $a = \alpha_i$ is open. By using the standard picture based on systems interacting with contexts, we can say that context C_{α_i} prepares systems corresponding to the value $a = \alpha_i$ of the a-observable.[5]

The condition (5.18) is the condition of *repeatability* presented in the abstract contextual setting. As we know repeatability is one of the basic features of classical observables represented by random variables (equation (1.27), section 1.5) and quantum observables of the von Neumann-Lüders type (equation (4.29), section 4.3). For these observables repeatability is derived from their basic properties. It can be treated as a mathematical theorem. In the abstract contextual model under consideration, repeatability is postulated. In principle, we can generalize our contextual model to non-repeatable observations and this is an interesting, although nontrivial, problem.

Generalized quantum observables given by POVM in general are not repeatable.[6] Thus our contextual model does not cover the case of POVM-observables.

We repeat once again that in our contextual model repeatability of observables is equivalent to existence of corresponding filtration contexts $C_\alpha, \alpha \in X_a$, where X_a is the range of values of a. These filtration contexts will play the crucial role in further considerations.

Under these assumptions, we obtain [161]:

$$P(b = \beta|C) = P(a = \alpha_1|C)P(b = \beta|C_{\alpha_1}) + P(a = \alpha_2|C)P(b = \beta|C_{\alpha_2})$$

$$+ 2\lambda_\beta \sqrt{P(a = \alpha_1|C)P(b = \beta|C_{\alpha_1})P(a = \alpha_2|C)P(b = \beta|C_{\alpha_2})} \tag{5.19}$$

---

[5]However, we try to proceed without systems. Hence, we prefer the picture of open $[a = \alpha_i]$-channel.

[6]The problem of repeatability for quantum observables is interesting and complicated. As was shown in [47], in finite dimensional state spaces repeatability for atomic instruments implies that they are of the von Neumann-Lüders type. However, in the infinite dimensional case this result is not valid, see again [47] for corresponding counterexamples. For non-atomic instruments, the problem has not yet been solved.

and the coefficient λ_i, *the interference coefficient*, is in general nonzero.

The derivation of FTP with the interference term is straightforward [161] and even tautological. Simply define the interference coefficients as

$$\lambda_\beta = \frac{P(b = \beta|C) - \sum_j P(a = \alpha_j|C)P(b = \beta|C_{\alpha_j})}{2\sqrt{P(a = \alpha_1|C)P(b = \beta|C_{\alpha_1})P(a = \alpha_2|C)P(b = \beta|C_{\alpha_2})}} \quad (5.20)$$

and then solve this equality with respect to the probability $P(b = \beta_i|C)$.

The meaning of the nominator is clear: it is the difference between two types of contextual probabilities, one can say "interference" between these contexts. The denominator gives a proper normalization, cf. section 5.3. In fact, my main observation [133], [156], [161] was just that in the absence of the set-theoretic representation of probabilities there are no reasons to expect that the nominator has to equal to zero.

5.3.3 *Non-Kolmogorovean probability models*

The coefficient λ can be used to test whether collected statistical data can be represented with the aid of the Kolmogorov probability model or not. If $\lambda \neq 0$ we can speak about *non-Kolmogorovean probability theory* [161]. As always in the case of impossibility statements, one has to take the statement about impossibility to use the classical (Kolmogorov) model with caution. There may exist a variety of ways to describe mathematically statistical data collected in experiments. It might be that the statement about impossibility is just a sign of "lack of imagination". By saying that nonzero coefficient of interference implies the absence of the Kolmogorov probability representation, we mean only that the straightforward representation through identification of frequencies of outcomes with probabilities with respect to *a single probability measure P* is impossible. However, this does not exclude that there might exist some more complex mathematical ways to embed data in the Kolmogorov space. It can even happen that some "not straightforward" mathematical representation matches better with the real experimental situation, see section 5.5.3 for the unconditional and conditional versions of (probabilistic) compatibility of observables. During the last 20 years I explored a lot the ideology of non-Kolmogorovean probability (stimulated by L. Accardi [2] - [4]), both in quantum physics and outside it, e.g., in economics, cognitive science, psychology, social science, see Chapter 9. However, recently I understood that in the light of the above argument about "lack of imagination", the use of the terminology *"non-Kolmogorovean probability theory"* and *"nonclassical probability theory"* might be ambiguous. Nevertheless, this terminology is useful to emphasize impossibility of the straigtforward application of the Kolmogorov model to some experimental statistical data. So, we shall use it (especially in Chapter 9), but with caution.

5.3.4   *Trigonometric and hyperbolic interference*

The coefficient of interference can be used as a measure of interference between incompatible observables a and b. If it happens (as in the two-slit experiment) that

$$|\lambda_\beta| \leq 1, i = 1, 2, \qquad (5.21)$$

we can introduce new parameters, "probabilistic phases", and represent the interference coefficients as

$$\lambda_\beta = \cos\theta_\beta. \qquad (5.22)$$

Then the generalized FTP (5.19) takes the form:

$$P(b = \beta|C) = P(a = \alpha_1|C)P(b = \beta|C_{\alpha_1}) + P(a = \alpha_2|C)P(b = \beta|C_{\alpha_2})$$

$$+ 2\cos\theta_\beta\sqrt{P(a = \alpha_1|C)P(b = \beta|C_{\alpha_1})P(a = \alpha_2|C)P(b = \beta|C_{\alpha_2})}. \quad (5.23)$$

However, in general *the interference coefficients λ_β determined by the equality (5.20) can exceed one!* We shall come back to this question at the very end of section 5.6.2; see also the monograph [161] for analysis of this case in the framework of so called *hyperbolic QM*.

Standard QM is algebraically based on complex numbers. Hyperbolic QM is based on so-called hyperbolic numbers, $z = x + jy, x, y \in \mathbf{R}$, and j satisfies the following squaring constraint $j^2 = 1$, cf. with $i^2 = -1$ for complex numbers. The set of hyperbolic numbers is a commutative algebra $\mathbf{G}$, but not a field, i.e., here we can add, subtract, multiply, but in general not divide. One of the main differences of hyperbolic Hilbert space from complex one is that the hyperbolic scalar product is not positive-definite. Thus hyperbolic QM is a kind of quantum theory with an indefinite scalar product. Here operating with probability amplitudes and corresponding probabilities is not so straightforward as in complex QM. For each observable, one has to select a class of states generating nonnegative probabilities.

Hyperbolic Hilbert space representation appears naturally starting with representation of interference terms exceeding 1 by using hyperbolic functions:

$$\lambda_\beta = \cosh \theta_\beta, \tag{5.24}$$

where $\theta_\beta \in \mathbf{R}$. This is a hyperbolic analog of phase. Then one proceeds similarly to section 5.4 devoted to reconstruction of complex probability amplitudes starting with FTP with the trigonometric interference (5.23). In the purely hyperbolic case, i.e., for both interference coefficients exceeding 1, FTP has the form:

$$P(b = \beta|C) = P(a = \alpha_1|C)P(b = \beta|C_{\alpha_1}) + P(a = \alpha_2|C)P(b = \beta|C_{\alpha_2})$$

$$+ 2\cosh \theta_\beta \sqrt{P(a = \alpha_1|C)P(b = \beta|C_{\alpha_1})P(a = \alpha_2|C)P(b = \beta|C_{\alpha_2})}. \tag{5.25}$$

The general situation is more complex. It can happen that one coefficient of interference exceeds 1 and another is bounded by one. Here one has to use a more general Hilbert space on the algebra of hyper-complex numbers [161].

5.4 Constructive Wave Function Approach

We recall that Born's rule can be interpreted as an algorithm to transfer complex amplitudes (or in the Hilbert space formalism – normalized state vectors) to probabilities. This rule was postulated by M. Born. It cannot be derived in the quantum formalism.

5.4.1 *Inverse Born's rule problem*

One can try to derive it by using other models of measurements. In this section we shall derive it from our contextual probabilistic model, see also [161]. The problem under study can be called *inverse Born's rule problem*:

IBP (inverse Born problem): *To construct a representation of probabilistic data by complex probability amplitudes (complex state vectors) that match Born's rule.*

Solution of IBP would provide a way to represent probabilistic data by "wave functions" ("waves of probabilities") and operate with this data using linear algebra (as we do in the conventional QM). In particular, one would be able to find quantum-like effects (such as interference of probabilities) in data collected in any domain of science. Such an activity can be called *constructive wave function approach.*

We remark that for quantum physics solution of IBP has merely the foundational value, clarification of the origin of Born's postulate. However, in recently flowering applications of the mathematical formalism of QM outside of physics, Chapter 9, solution of IBP is very important for applications. By solving IBP we construct the quantum-like complex Hilbert representation of probabilistic data. In contrast to physics, in, e.g., cognitive science and psychology we do not have the quantization procedure similar to Schrödinger's quantization. The situation is even worse. Here we do not have even an analog of the classical physical representation, e.g., in the form of phase space mechanics. In fact, there is nothing to quantize. Therefore here we have to start directly from operational models of experiment.

As was already pointed out, in this section we plan to start with our contextual probabilistic model. However, before proceeding in this way, we briefly analyze another possibility is derive Born's rule, namely, from some subquantum model. This approach is interesting not only for physics, where classical field theory played the crucial role in QM: De Broglie, Schrödinger, Einstein, Bohm,..., but even for brain studies and cognition, where the quantum-like structure of information processing by the brain might (but need not!) be coupled to classical electromagnetic fields generated by it, [154].

Thus let us compare Born's rule of QM with models of the classical field type. Here Born's rule appears naturally: higher field intensity leads to higher probability of detection and field's energy is given by the squaring of its amplitude. For example, consider very generally, i.e., without coupling to any concrete subquantum model, the complex representation of the classical electromagnetic field $\phi(x) = (E(x), B(x))$. This is the *Riemann-Silberstein representation:*

$$\Phi(x) = E(x) + iB(x). \tag{5.26}$$

In this representation field's energy is given by the square of the absolute value of the complex field:

$$\text{Energy}(x) \equiv E^2(x) + B^2(x) = |\Phi(x)|^2. \tag{5.27}$$

Finiteness of field's total energy has the form of the square integrability condition for the complex field:

$$\text{Energy} \equiv \int (E^2(x) + B^2(x))dx = \int |\Phi(x)|^2 dx < \infty. \tag{5.28}$$

Since a click of a detector interacting with the electromagnetic field and located at the point x is generated by supply of field's energy to this point, we obtain the following version of Born's rule:

$$p(x) = \frac{\text{Energy}(x)}{\text{Energy}} = \frac{|\Phi(x)|^2}{\int |\Phi(x)|^2 dx} = |\psi(x)|^2, \qquad (5.29)$$

where the normalized (by the total energy) electromagnetic field is defined as

$$\psi(x) = \frac{\Phi(x)}{\sqrt{\int |\Phi(x)|^2 dx}}. \qquad (5.30)$$

By following this line of reasoning it is possible to elaborate models of *threshold detection* of the classical electromagnetic field reproducing quantum probabilities for photon counting. In particular, in the rigorous mathematical way it was done [167] in the framework of *prequantum classical statistical field theory*, a random field model generating the quantum representation of states, observables and probabilities (section 5.2.2). However, in PCSFT approach Born's rule appeared not as the exact rule, not as a law of nature (cf. with QBism, section 6.7.3), but as an *approximative rule for calculation of detection probabilities*.

It seems that classical field (wave) theory also played the crucial role in motivating Schrödinger to introduce the wave function (section 5.2). For him, the squared wave function gives the density of the electron charge, see (5.4). It is clear that the probability to detect electron is higher in domains where density is higher. Thus identification of the squared field with charge's density $r(x)$ leads to Schrödinger's version of Born's rule for the probability of finding an electron at the point x :

$$p(x) = \frac{r(x)}{\int r(x) dx} = |\psi(x)|^2, \qquad (5.31)$$

where the normalization $\psi(x)$ of Schrödinger's wave function $\Phi(x)$ is given by (5.30). Here $\Phi(x)$ is the amplitude of the density of the electron charge which is not normalized.

In some sense the aforementioned classical field approach to IBP is trivial (at the level of ideas; technical derivations can be very complicated [167]). Here we start directly from the complex Hilbert space and the squared absolute value is the simplest quantity having the physical meaning, cf. with the discussion in section 6.7.5. We recall that the L^2-space is widely used in the classical field theory and the complex (Riemann-Silberstein) representation is standard in radio-engineering and signal processing. In fact, QM borrowed the L^2 state space from the classical field theory.

5.4.2　*Quantum-like representation algorithm*

Solution of IBP starting with an abstract contextual probabilistic model of measurements is ideologically more complicated. From the very beginning, there is no complex Hilbert space nor even complex numbers. Probabilities are given by real numbers. First one has to find some fingerprint of complex representation in calculus of real probabilities. For me, such a fingerprint is the presence of the cosines in the contextual version of FTP, see (5.23). Then I coupled it with Euler's formula for complex numbers [161].

After such a long introduction, now we go to technical details. We first shorten notations for contextual probabilities, by setting:

$$p_C^b(\beta) \equiv P(b = \beta|C), p_C^a(\alpha) \equiv P(a = \alpha|C), p_{\beta|\alpha} \equiv P(b = \beta|C_\alpha).$$

We call probabilities $p_{\beta|\alpha}$ *transition probabilities*, cf. with transition probabilities in QM (section 4.3).

The idea is to represent some class of contexts with the aid of probabilities for outputs of measurement for two observables a and b, *reference observables* in terminology used in [161]. Such a pair of abstract observables mimics a pair of conjugate variables in classical mechanics or observables in QM, such as *position and momentum*. In physics the notions of conjugate variables (classical) and observables (quantum) are expressed in mechanical terms, as nontrivial Poisson brackets and commutators, respectively. However, such formulations presume the existence of a linear space representation.

We do not start with this representation, but we want to derive it. Therefore we have to formulate a purely probabilistic analog of conjugacy of variables/observables. We start with the remark that in QM conjugacy of observables is closely coupled to the notion of *complementarity of observables*. We shall explore a notion similar to complementarity, but not coinciding with it. We call it *supplementarity*, i.e., we shall represent contexts in complex Hilbert space with the aid of a pair of *supplementary observables*. This notion was invented and intensively explored in [153]. As well as Bohr's complementarity, it reflects the impossibility of joint precise determination. However, in contrast to Bohr, I do not exclude the possibility to model such observables in the realist framework, see section 6.1.1 on (non)realist interpretations of QM, see also section 6.9 on the Växjö interpretation of QM (the realist contextual interpretation).

However, first we proceed formally without formulating explicitly restrictions on observables and contexts.

We rewrite FTP with *the interference term of the trigonometric type* (and in this book we are interested only in such type of interference, see (5.23)), in shorten notations:

$$p_C^b(\beta) = p_C^a(\alpha_1)p_{\beta|\alpha_1} + p_C^a(\alpha_2)p_{\beta|\alpha_2}$$

$$+ 2\cos\theta_\beta \sqrt{p_C^a(\alpha_1)p_{\beta|\alpha_1}p_C^a(\alpha_2)p_{\beta|\alpha_2}}. \quad (5.32)$$

We now use the elementary formula

$$D = A + B + 2\sqrt{AB}\cos\phi = |\sqrt{A} + e^{i\phi}\sqrt{B}|^2,$$

for real numbers $A, B > 0, \phi \in [0, 2\pi]$ to introduce a complex amplitude corresponding to FTP (5.32):

$$\psi(\beta) \equiv \psi_C(\beta) = \sqrt{p_C^a(\alpha_1)p_{\beta|\alpha_1}} + e^{i\phi_\beta}\sqrt{p_C^a(\alpha_2)p_{\beta|\alpha_2}}, \quad \beta \in X_b. \quad (5.33)$$

This amplitude can be used to represent the probability $p_C^b(\beta)$ as the square of the complex amplitude (Born's rule):

$$p_C^b(\beta) = |\psi(\beta)|^2. \quad (5.34)$$

The formula (5.33) represents the *quantum-like representation algorithm* (QLRA) [141]. For any "trigonometric context" C, QLRA produces the complex amplitude ψ. This algorithm can be used in any domain of science to create the quantum-like (complex Hilbert space) representation of probabilistic data.

We denote the space of functions $\psi : X_b \to \mathbf{C}$ by the symbol $\Phi = \Phi(X_b, \mathbf{C})$. Since $X_b = \{\beta_1, \beta_2\}$, Φ is the two-dimensional complex linear space.

By the symbol C^{Tr} we denote the class of contexts for which both coefficients of interference

$$\lambda_\beta \equiv \frac{p_C^b(\beta) - (p_C^a(\alpha_1)p_{\beta|\alpha_1} + p_C^a(\alpha_2)p_{\beta|\alpha_2})}{2\sqrt{p_C^a(\alpha_1)p_{\beta|\alpha_1}p_C^a(\alpha_2)p_{\beta|\alpha_2}}} \quad (5.35)$$

in generalized FTP (5.32) are bounded by 1. Of course, this class of contexts depends on the pair of reference observables a, b, i.e., $C^{\mathrm{Tr}} = C^{\mathrm{Tr}}(b|a)$. In fact, condition (5.35) contains a number of constraints on contextual probabilities; later we shall analyze them in more detail.

QLRA determines the map

$$J^{b|a} : C^{\mathrm{Tr}} \to \Phi(X, \mathbf{C}).$$

It maps probabilistic data about contexts (of the special class) collected with the aid of two concrete reference observables into complex amplitudes.

The representation (5.34) of probability is nothing but the famous *Born rule*. The complex amplitude $\psi(\beta)$ can be called a *wave function of context C* in the (a, b) representation. More precisely we have to speak about $b|a$ representation, since it is based on transition probabilities from the a-observation to the b-observation.

We set

$$e_\beta^b(x) = \delta(\beta - x), \ x \in X_b,$$

Dirac delta-functions concentrated in points $\beta = \beta_1, \beta_2$, i.e., $e_\beta^b(x) = 1$ if and only if $x = \beta$.

Born's rule for complex amplitudes (5.34) can be rewritten in the following form:

$$p_C^b(\beta) = |\langle \psi, e_\beta^b \rangle|^2,$$

where the scalar product in the space $\Phi(X_b, \mathbf{C})$ is defined by the standard formula

$$\langle \psi_1, \psi_2 \rangle = \sum_{\beta \in X_b} \psi_1(\beta)\overline{\psi_2(\beta)}. \tag{5.36}$$

The system of functions $\{e_\beta^b\}_{\beta \in X_b}$ is an orthonormal basis in the Hilbert space $H = (\Phi, \langle \cdot, \cdot \rangle)$:

$$e_{\beta_1}^a = \begin{pmatrix} 1 \\ 0 \end{pmatrix}, \quad e_{\beta_2}^a = \begin{pmatrix} 0 \\ 1 \end{pmatrix}. \tag{5.37}$$

Now let $X_b \subset \mathbf{R}$ (in general β is just a label for a result of observation). By using the Hilbert space representation of Born's rule we obtain the Hilbert space representation of the expectation of the observable b,

$$E[b|C] = \sum_{\beta \in X_b} \beta|\psi_C(\beta)|^2 = \sum_{\beta \in X_b} \beta\langle \psi_C, e_\beta^b \rangle\overline{\langle \psi_C, e_\beta^b \rangle} = \langle B\psi_C, \psi_C \rangle,$$

$$\tag{5.38}$$

where the Hermitian operator $B : H \to H$ is determined by its eigenvectors: $Be_\beta^b = \beta e_\beta^b, \beta \in X_b$. This is the multiplication operator in the space of complex functions $\Phi(X_b, \mathbf{C}) : B\psi(\beta) = \beta\psi(\beta)$. It is natural to represent the b-observable (in the Hilbert space model) by the diagonal operator

$$B = \begin{pmatrix} \beta_1 & 0 \\ 0 & \beta_2 \end{pmatrix}. \tag{5.39}$$

We remark that previous considerations can be applied to any matrix of transition probabilities $\mathbf{P}^{b|a} = (p_{\beta|\alpha})$.

To solve IBP completely, we would like to have Born's rule not only for the b-variable, but also for the a-variable:

$$p_C^a(\alpha) = |\langle \psi, e_\alpha^a \rangle|^2 \, , \alpha \in X_a.$$

How can we define the basis $\{e_\alpha^a\}$ corresponding to the a-observable?

This basis can also be found by starting with interference of probabilities and the corresponding complex amplitude given by (5.33). We have

$$\psi = \sqrt{p_C^a(\alpha_1)} f_{\alpha_1}^a + \sqrt{p_C^a(\alpha_2)} f_{\alpha_2}^a \, , \tag{5.40}$$

where

$$f_{\alpha_1}^a = \begin{pmatrix} \sqrt{p_{\beta_1|\alpha_1}} \\ \sqrt{p_{\beta_2|\alpha_1}} \end{pmatrix}, \quad f_{\alpha_2}^a = \begin{pmatrix} e^{i\phi_{\beta_1}} \sqrt{p_{\beta_1|\alpha_2}} \\ e^{i\phi_{\beta_2}} \sqrt{p_{\beta_2|\alpha_2}} \end{pmatrix}. \tag{5.41}$$

However, this system of vectors can be orthonormal only if the matrix of transition probabilities $\mathbf{P}^{b|a}$ is *doubly stochastic*.

5.4.3 *Double stochasticity*

We recall the definition of double stochasticity, section 4.3:

$$\sum_\beta p_{\beta|\alpha} = 1, \quad \sum_\alpha p_{\beta|\alpha} = 1. \tag{5.42}$$

We have already seen, section 4.3, that in Hilbert space of any dimension pairs of von Neumann-Lüders observables with discrete non-degenerate spectra produce doubly stochastic matrices of transition probabilities. Below we shall show that in the case of dichotomous observables (of any origin) double stochasticity induces their representation as von Neumann-Lüders observables, Hermitian operators, in (two dimensional) Hilbert space.

We remark that for non-dichotomous observables double stochasticity of the matrix of transition probabilities is only the necessary condition, but not sufficient to represent them by Hermitian operators in complex Hilbert space. Already in the case of three-valued observables, we are not able to characterize the class of matrices which can be represented as matrices of transition probabilities for pairs of von Neumann-Lüders observables, see [207], [208] for an attempts to proceed in this direction.

Further we proceed with dichotomous observables with the double stochastic matrix of transition probabilities.

Double stochasticity implies that the system of vectors $\{f_{\alpha_i}^a\}$ is an orthonormal basis iff the probabilistic phases satisfy the constraint (see [161], [163])

$$\phi_{\beta_2} - \phi_{\beta_1} = \pi \bmod 2\pi \, , \tag{5.43}$$

i.e., the phases cannot be chosen independently. Thus, instead of the a-basis (5.41), which depends on phases, we can consider a new a-basis that depends only on the matrix of transition probabilities $\mathbf{P}^{b|a}$

$$e^a_{\alpha_1} = \begin{pmatrix} \sqrt{p_{\beta_1|\alpha_1}} \\ \sqrt{p_{\beta_2|\alpha_1}} \end{pmatrix}, \qquad e^a_{\alpha_2} = \begin{pmatrix} \sqrt{p_{\beta_1|\alpha_2}} \\ -\sqrt{p_{\beta_2|\alpha_2}} \end{pmatrix}. \tag{5.44}$$

In this basis ψ is represented as

$$\psi = \sqrt{p^a_C(\alpha_1)}e_{\alpha_1} + e^{i\phi_{\beta_1}}\sqrt{p^a_C(\alpha_2)}e_{\alpha_2}. \tag{5.45}$$

The a-observable is represented by the Hermitian operator A, which is diagonal with eigenvalues α_1, α_2 in the basis $\{e^a_\alpha\}$. The average of the observable a coincides with the quantum Hilbert space average:

$$E[a|C] = \sum_{\alpha \in X_a} \alpha p^a_C(\alpha) = \langle A\psi_C, \psi_C \rangle. \tag{5.46}$$

We recall that the matrix of transition probabilities $\mathbf{P}^{b|a}$ was assumed to be doubly stochastic. Thus

$$e^a_{\alpha_1} = \begin{pmatrix} \sqrt{p} \\ \sqrt{1-p} \end{pmatrix}, \qquad e^a_{\alpha_2} = \begin{pmatrix} \sqrt{1-p} \\ -\sqrt{p} \end{pmatrix}. \tag{5.47}$$

Here $p = p_{\beta_1|\alpha_1} = p_{\beta_2|\alpha_2}$ (this is a trivial consequence of double stochasticity). In the basis $\{e^b_\beta\}$ the operator A is represented by the matrix:

$$A = \begin{pmatrix} \alpha_1 p + \alpha_2(1-p) & (\alpha_1 - \alpha_2)\sqrt{p(1-p)} \\ (\alpha_1 - \alpha_2)\sqrt{p(1-p)} & \alpha_1(1-p) + \alpha_2 p \end{pmatrix}. \tag{5.48}$$

5.4.4 *Supplementary observables*

We turn to the general case, i.e., without the double stochasticity constraint, and discuss constraints incorporated in the definition of the coefficient of interference (5.35). This was promised when we introduced the class of trigonometric contexts $\mathcal{C}^{\mathrm{Tr}}(b|a)$. It is clear that quantities in the denominator of the expression (5.35) have to be nonzero. There are probabilities of two different types: a) the a-probabilities with respect to the context C; b) the transition probabilities.

A context C is said to be a-*nondegenerate* if

$$p^a_C(\alpha) \neq 0 \tag{5.49}$$

for any value α of a. Thus elements of $\mathcal{C}^{\mathrm{Tr}}(b|a)$ have to be a-nondegenerate.

Observables a and b are called $[b|a]$-supplementary if

$$p_{\beta|\alpha} = P(b = \beta|C_\alpha) \neq 0, \ \beta \in X_b, \alpha \in X_a. \tag{5.50}$$

We remark that context C_α was determined by the condition (5.18), i.e., conditioning with respect to context C_α corresponds to conditioning with respect to the value $a = \alpha$ of the a-observable. The constraint (5.52) can be rewritten as

$$p_{\beta|\alpha} = P(b = \beta|C_\alpha) \neq 1, \ \beta \in X_b, \alpha \in X_a. \tag{5.51}$$

Thus determination of the value of a with probability 1, $a = \alpha$, implies impossibility to determine the value of b with probability 1. This is the purely probabilistic version of the principle of complementarity of N. Bohr. However, as was already pointed out, we do not want to speak about "complementarity", since this notion had already been reserved by N. Bohr and it presumes mutual exclusivity - incompatibility, impossibility of joint measurement, the absence of the joint probability distribution of the pair of observables, $p^{a,b}(\alpha, \beta)$, see section 5.5.3 on observational and probabilistic incompatibility. Supplementarity is a weaker restriction on inter-relation of observables. It does not mean mutual exclusivity; in principle supplementary observables can be observationally and probabilistically compatible.

The latter means that they can be represented as random variables on the same Kolmogorov probability space $\mathcal{P} = (\Omega, \mathcal{F}, P)$. In such a case, transition probabilities are given by Bayes' formula

$$P(b = \beta|C_\alpha) = P(a = \alpha, b = \beta)/P(a = \alpha).$$

Supplementarity is reduced to the restriction

$$P(a = \alpha, b = \beta) \neq 0, \ \beta \in X_b, \alpha \in X_a. \tag{5.52}$$

In the same way we can define $[a|b]$-supplementary if

$$p_{\alpha|\beta} = P(a = \alpha|C_\beta) \neq 0, \ \beta \in X_b, \alpha \in X_a, \tag{5.53}$$

where contexts $C_\beta \equiv C_\beta^b$ are determined as filtration contexts with respect to the values of the b-observable. This is contextual formalization of *repeatability* of the b-observable:

$$P(b = \beta_i|C_{\beta_i}) = 1, \tag{5.54}$$

$$P(b = \beta_i|C_{\beta_j}) = 0, i \neq j. \tag{5.55}$$

We call observables *supplementary* if they are both $[b|a]$- and $[a|b]$-supplementary. We remark that if observables are probabilistically compatible, see (5.52), then the notions $[a|b]$- and $[a|b]$-supplementarities coincide, because it is reduced to the constraint (5.52) which is symmetric with respect to $[a|b]$- and $[a|b]$-conditionings.

We showed that any pair of $[b|a]$-supplementary dichotomous observables with double stochastic matrix of transition probabilities can be represented by Hermitian operators in complex Hilbert space. It is easy to check (by using their matrix presentations) that *these operators do not commute.* Thus from the quantum mechanical viewpoint these observables are incompatible, i.e., they cannot be measured jointly. In any event the mathematical formalism of QM does not determine their joint probability distribution. However, as was pointed out, these observables can be probabilistically compatible, i.e., they can have the joint probability distribution and be represented by random variables on classical probability space.

5.4.5 *Symmetrically conditioned observables*

Two observables are called *symmetrically conditioned* if

$$p_{\beta|\alpha} = p_{\alpha|\beta}, \beta \in X_b, \alpha \in X_a, \tag{5.56}$$

or in more detailed notations:

$$P(b = \beta|C_\alpha) = P(a = \alpha|C_\beta). \tag{5.57}$$

We remark that this condition implies double stochasticity.

However, in contrast with the Kolmogorov case, in our general contextual model double stochasticity even of both matrices of transition probabilities $\mathbf{P}^{b|a}$ and $\mathbf{P}^{a|b}$ does not imply symmetric conditioning. We recall that in QM any two observables of the von Neumann-Lüders type with nondegenerate spectra are symmetrically conditioned (for the state space of an arbitrary dimension, section 4.3). It is a consequence of the symmetry of the scalar product.

In our approach only observables a and b (reference observables) were fixed from the very beginning. The operator representation of these observables, $a \to A, b \to B$, was constructed on the purely probabilistic basis (by using QLRA).

Suppose now that we have a third observable, say c. We can couple it either with a or with b. In the first case we proceed with QLRA for the pair of reference observables (c, a) and in the second case (c, b). We obtain two representations in complex Hilbert space.

Question: *Under which conditions are these representations unitary equivalent?*

Answer [161]: *If for all pairs of observables the matrices of transition probabilities are symmetrically conditioned.*

In the class of pairwise symmetrically conditioned observables all Hilbert space representations of them (in particular, $b|a$ and $a|b$ representations) are unitary equivalent. Thus we recovered this basic feature of QM, the unitary equivalence of different representations, starting from the purely contextual probabilistic model.

5.4.6 *Non-doubly stochastic matrices of transition probabilities*

Consider a pair of the reference (dichotomous) observables (a, b), i.e., those used to generate the QLRA-representation such that the matrix of transition probabilities $\mathbf{P}^{b|a}$ is not doubly stochastic. If these observables are $[b|a]$-supplementary, then for trigonometric contexts the QLRA-algorithm works well. The b-observable can be represented by the diagonal Hermitian operator B with matrix (5.39) with respect to the canonical basis. However, the operator A given by the matrix (5.48) which is not Hermitian. Thus a cannot be represented by QLRA as the von Neumann-Lüders observable. May be it can be represented as POVM? Unfortunately, in general the answer is no. The operator A with the matrix (5.48) can be treated as a "non-normalized POVM", i.e., $M = (M_1, M_2), M_1 + M_2 \neq I$, see [161].

In applications outside of physics (Chapter 9), e.g., cognition and psychology, *the matrices of transition probabilities calculated with the aid of experimental probabilistic data are always not doubly stochastic* [163]. (It is not clear why this happens.) Thus straightforward application of QLRA to such data is impossible. One has to work with non-doubly stochastic matrices of transition probabilities and to use generalizations of POVMs.

5.5 Contextual Probabilistic Description of Measurements

5.5.1 *Contexts, observables, and measurements*

A fundamental notion of this model is *context*. It is a complex of conditions, e.g., physical, or biological, or mental, or financial, or political. Another fundamental notion is *observable*. Construction of our contextual probabilistic model starts with selection of the families of contexts $\mathcal{C}$ and observables O.

Denote observables by Latin letters, $a, b, ...$, and their values by Greek letters, $\alpha, \beta, ...$. For an observable a, denote the set of its possible values by the symbol X_a. To simplify considerations, we will consider only discrete observables.[7]

The next step is selection, for each context $C \in \mathcal{C}$, of a family of *observables* $O_C \subset O$ which can be measured for this context. It is important to remark that $b_1, ..., b_k \in O_C$ need not be *"observationally compatible"*, i.e., the possibility of their joint measurement is not guaranteed; in other words the vector observable $b = (b_1, ..., b_k)$ need not belong to O_C. We also remark that all sets $O_C, C \in \mathcal{C}$, are considered as subset of the O, the total family of observables of the model. This is useful, since we can speak about "an observable" even irrelative to concrete context, as an element of O. By concreting context, by saying that $b \in O_c$, we point to the special contextual realization of b. (It might even be natural to label such contextual realization of b by the symbol b_C, but by simplicity reason we shall not do this; see also below the definition of measurement.)

Denote by the symbol $\mathcal{O}_{\mathcal{C}}$ the collection of families of observables corresponding to contexts belonging $\mathcal{C}$, namely $\mathcal{O}_{\mathcal{C}} = \{O_C\}_{C \in \mathcal{C}}$.

In this framework *measurement* is represented by a pair (C, b), where $b \in O_C$. We remark that the same observable b can belong to O_{C_1} and O_{C_2}, where C_1 differs from C_2. The pairs (C_1, b) and (C_2, b) represent two different measurements, observations of b under contexts C_1 and C_2, respectively.

Example 1. Consider the two-slit experiment, section 5.1, and contexts $C_i, i = 0, 1$, only the ith slit is open, and C_{01} both slits are open. Consider the observable b, determination of momentum through detection of the point on the photo-emulsion screen. Then this observable can be measured in three different contexts: $(C_i, b), i = 0, 1, (C_{01}, b)$. Consider also the observable a, determination of position through "which slit detection." Then we again have three different measurements $(C_i, a), i = 0, 1, (C_{01}, a)$. In measurement (C_i, a) the observable $a = i$ with probability one. In the measurement (C_{01}, a) (and symmetric location of the source with respect to the slits) $a = i, i = 0, 1$, with probability $1/2$. We remark that, for all contexts, C_i, C_{01}, observables a and b are not observationally compatible, it is impossible to measure them jointly.

This example shows that the notion of context differs from the notion of

---

[7] We remark that our general model does not contain systems. There are only observables and context under which observations can be performed.

a *preparation procedure* which is widely used in the operational (empiricist) approach to QM [71]. In the two-slit experiment the preparation procedure represents the source, e.g., a source of electromagnetic field (classical or quantum light). Context is not reduced to the representation of the source. It represents also the screen with slits. Roughly speaking in general we extend the area of preparation and reduce the area of observation. The momentum observable b is represented by the photo-emulsion screen; the position observable a is represented by detectors located after slits.

Example 2. Consider the EPR-Bohm-Bell experiment (see Chapter 8) and contexts $C_{\theta\gamma}$, where θ and γ are orientations of polarization beam splitters (PBSs), on "the left-hand ans right-hand sides", respectively. Consider observables a_θ and b_γ - projections of photon polarization on the axes determined by the angles θ and γ. Then, e.g., for the observable a_θ, we have a continuous family of measurements labeled by the angle γ, $(C_{\theta\gamma}, a_\theta)$. Moreover, for each context $C_{\theta\gamma}$, observables a_θ and b_γ are compatible. Hence, joint measurements $(C_{\theta\gamma}, (a_\theta, b_\gamma))$ are well defined. Consider contexts $C_\theta^{(L)}$, the presence of the θ-oriented PBS only on the left-hand side; in the same way we define contexts $C_\gamma^{(R)}$. The polarization observables can also be measured in these contexts: $(C_\theta^{(L)}, a_\theta), (C_\gamma^{(R)}, b_\gamma)$.

We remark again that aforementioned contexts are not reduced to the commonly used preparation procedure representing a source of entangled photons. For example, context $C_{\theta\gamma}$ represents not only the source, but also two PBSs oriented with angles θ and γ, respectively. Observables are represented by pairs of detectors located at outputs channels of corresponding PBSs.

5.5.2 Contextual probabilistic model

Definition 1. [161] *A contextual probabilistic model is a triple*

$$\mathcal{P}_C = (\mathcal{C}, \mathcal{O}_C, \pi),$$

where elements of $\mathcal{C}$ and $\mathcal{O}_C$ are contexts and observables, and elements of π are the corresponding probability distributions satisfying the conditions of consistency, see below (5.59), (5.61).

Here $\pi = \{p_C^a\}, C \in \mathcal{C}, a \in \mathcal{O}_C$, and, for any $\alpha \in X_a$,

$$p_C^a(\alpha) \equiv P(a = \alpha | C) \tag{5.58}$$

is the probability of the value $a = \alpha$ for observation of a under context C.

We assume that p_C^a is a probability measure.[8] In particular, we have (as the result of using the measure theory) that

$$\sum_{\alpha \in X_a} p_C^a(\alpha) = 1.$$

Since we consider only discrete observables, we can select the σ-algebra of all subsets of X_a as collection of events for measurements of a. Then by definition

$$p_C^a(D) = \sum_{\alpha \in X_a} p_C^a(\alpha)$$

for any subset D of X_a, In particular, $p_C^a(X_a) = 1$.

Now we specify the conditions of consistency which were mentioned in Definition 1, cf. with the corresponding conditions of the Kolmogorov theorem about stochastic processes, section 1.9.4. Suppose that observables $b_1, ..., b_m \in C$ are compatible with respect to this context, i.e., the vector-observable $b = (b_1, ..., b_m) \in C$. Then, the joint probability distribution is well defined, $p_C^{(b_1,...,b_m)}(\beta_1, ..., \beta_m) \equiv p_C^b(\beta_1, ..., \beta_m)$. It is natural to assume (and we do this in Definition 1) that probabilities with respect to context C satisfy the following condition of consistency of probability distributions with respect to fixed context C : for each b_i,

$$p_C^{b_i}(\beta_i) = \sum_{j \neq i} \sum_{\beta_j \in X_{b_j}} p_C^{(b_1,...,b_i,...,b_m)}(\beta_1, ..., \beta_i, ..., \beta_m). \tag{5.59}$$

This is the direct analog of condition (1.45), Chapter 1 (applied to discrete variables). It can also be written as

$$p_C^{b_i}(\beta_i) = p_C^{(b_1,...,b_i,...,b_m)}(X_{b_1}, ..., \beta_i, ..., X_{b_m}), \tag{5.60}$$

where X_{b_j} is the rannge of values of b_j, cf. again with (1.45).

The analog of condition (1.44), Chapter 1, can be formulated as follows. For any permutation $b_{i_1}, ..., b_{i_m}$ of compatible observables $b_1, ..., b_m \in C$,

$$p_C^{(b_{i_1},...,b_{i_m})}(\beta_{i_1}, ..., \beta_{i_m}) = p_C^{(b_1,...,b_m)}(\beta_1, ..., \beta_m). \tag{5.61}$$

The condition of consistency of probability distributions (5.59) has the following trivial, but important for applications, consequence. Let an observable a be compatible with two groups of observables, $b_1, ..., b_m$ and

---

[8] *Contextual expectation* $E[a|C]$ of an observable $a \in \mathcal{O}$ with respect to context $C \in \mathcal{C}$ is given by $\bar{a}_C = E[a|C] = \sum_{\alpha \in X_a} \alpha \, p_C^a(\alpha)$; contextual dispersion $D[a|C] = \sum_{\alpha \in X_a} (\alpha - \bar{a}_C)^2 \, p_C^a(\alpha)$.

$c_1, ..., c_k$, i.e., the vector-observables $(a, b_1, ..., b_m)$ and $(a, c_1, ..., c_k)$ belong to O_C, then

$$\sum_j \sum_{\beta_j \in X_{b_j}} p_C^{(a,b_1,...,b_m)}(\alpha, \beta_1, ..., \beta_m) = \sum_j \sum_{\mu_j \in X_{c_j}} p_C^{(a,c_1,...,c_k)}(\alpha, \mu_1, ..., \mu_k),$$

(5.62)

since both sides are equal to $p_C^a(\alpha)$ as a consequence of (5.59). We call the equality (5.62) the condition of *marginal consistency* with respect to fixed context C. It can also be written as

$$p_C^{(a,b_1,...,b_m)}(\alpha, X_{b_1}, ..., X_{b_m}) = p_C^{(a,c_1,...,c_k)}(\alpha, X_{c_1}, ..., X_{c_k}),$$

(5.63)

cf. (5.60).

We call probabilities (5.58) *contextual probabilities*. We may call them conditional probabilities as in Kolmogorov's model. Unlike the latter, contextual probability (5.58) is not the probability that an event, say B, occurs under the condition that another event, say C, occurred. Contextual probability is the probability of the result $a = \alpha$ under context C.

Although the consistency conditions (5.59), (5.61) naturally generalize the consistency conditions of the Kolmogorov theorem (1.44), (1.45), they do not lead to the Kolmogorov probability space representation of contextual probabilities for concrete context C. The pitfall is that in general O_C can contain incompatible observables, as in the above contextual probabilistic model for the two-slit experiment. More general problem is incompatibility of observables with respect to different contexts.

Consider the contextual probabilistic model corresponding to the EPR-Bohm-Bell experiment, Example 2. We proceed with the family of contexts $\mathcal{C} = \{C_{\theta\gamma}\}$ and family of observables a_θ, b_γ and vector-observables (a_θ, b_γ). Probabilities $p_{C_{\theta\gamma}}^{(a_\theta, b_\gamma)}$ are given by the quantum formalism. Then the consistency condition (5.59) has the form:

$$p_{C_{\theta\gamma}}^{a_\theta}(\alpha) = \sum_\beta p_{C_{\theta\gamma}}^{(a_\theta, b_\gamma)}(\alpha, \beta).$$

(5.64)

Of course, it would be interesting to compare probabilities $p_{C_{\theta\gamma_1}}^{a_\theta}(\alpha)$ and $p_{C_{\theta\gamma_2}}^{a_\theta}(\alpha)$. However, the corresponding measurements $(C_{\theta\gamma_1}, a_\theta)$ and $(C_{\theta\gamma_2}, a_\theta)$ are related to different contexts. Therefore it is useful to generalize the consistency conditions in the inter-contextual framework. From the very beginning we stress that it will be purely mathematical generalization; it relevance to physics (or biology) has to be clarified for each concrete contextual model.

5.5.3   *Probabilistic compatibility (noncontextuality)*

5.5.3.1   *Unconditional compatibility*

Consider a contextual probability model $\mathcal{P}_C = (\mathcal{C}, \mathcal{O}_C, \pi)$. We are now interested in (im)possibility to represent some family of contexts of this model with the aid of single Kolmogorov probability space.

Definition 2. *A family of contexts $\tilde{\mathcal{C}} = \{C_i\}$ belonging $\mathcal{C}$ is called (unconditionally) probabilistically compatible if there exists Kolmogorov probability space $\mathcal{P} = (\Omega, \mathcal{F}, P)$ such that all observables with respect to these contexts can be represented by random variables, i.e., there exists a map*

$$j : \cup_i \mathcal{O}_{C_i} \to R(\mathcal{P}), \tag{5.65}$$

such that, for each C_i and $b \in \mathcal{O}_{C_i}$,

$$p^b_{C_i}(\beta) = P(\omega \in \Omega : j(b)(\omega) = \beta).$$

Here, as in Chapter 1, the symbol $R(\mathcal{P})$ denotes the space of random variables for the probability space $\mathcal{P}$. We also remark that each set of observables $\mathcal{O}_{C_i}$ is a subset of the complete set of observables O of the contextual probabilistic model $\mathcal{P}_C$. Thus if $b \in \mathcal{O}_{C_k}$ and $b \in \mathcal{O}_{C_j}$, then $j(b) \in R(\mathcal{P})$ does not depend on contexts. Therefore we can also say that such a family of contexts $\tilde{\mathcal{C}}$ permits *noncontextual probabilistic description* or that the Kolmogorov probability space $\mathcal{P}$ from Definition 2 is a noncontextual probabilistic model for the family of contexts $\tilde{\mathcal{C}}$ (and the corresponding family of observables).[9]

If a family of contexts is probabilistically compatible, then, for any finite group of observables $b_i \in \mathcal{O}_{C_i}, i = 1, ..., m$, the corresponding group of random variables has the joint probability distribution given by

$$p^{b_1 \cdots b_m}(\beta_1, ..., \beta_m) \equiv P(\omega \in \Omega : j(b_1)(\omega) = \beta_1, ..., j(b_m)(\omega) = \beta_m). \tag{5.66}$$

Consider one fixed context C, here, for any $b_1, ..., b_m \in \mathcal{O}_C$ (observationally compatible or not), the joint probability distribution, see (5.66), is also well defined. If these observables are observationally compatible, then $p^{b_1 \cdots b_m}(\beta_1, ..., \beta_m) = p_C^{b_1 \cdots b_m}(\beta_1, ..., \beta_m)$.

We point out that probabilistic compatibility is a purely mathematical version of compatibility. One should distinguish it from *observational*

---

[9] Thus *probabilistic incompatibility is synonymous of noncontextuality*. The latter notion is better recognized for experts in quantum foundations. However, we prefer the terminology of Definition 2, since even in the case of probabilistic compatibility contexts are nontrivially present in the model.

compatibility, i.e., the possibility to perform joint measurement of these observables under some common context C. We remark that in general Kolmogorovian representation (for probabilistically compatible contexts) is not unique. Therefore the probability distributions $p^{b_1 \cdots b_m}(\beta_1, ..., \beta_m)$ are representation dependent. We remark that existence of the measure-theoretic representation implies the following condition of *marginal consistency*:

$$p_{C_i}^{b_i}(\beta_i) = \sum_{j \neq i} \sum_{\beta_j} p^{b_1 \cdots b_m}(\beta_1, ..., b_i, ..., \beta_m). \tag{5.67}$$

It can be written in the form analogous to (5.62). We restrict considerations to the case of two contexts.

Suppose that two contexts C_1 and C_2 are probabilistically compatible. Let an observable $a \in O_{C_i}, i = 1, 2$. Let $b_i \in O_{C_i}$ and the vector-observables $(a, b_i) \in O_{C_i}$, i.e., a is compatible with b_i for the context C_i. There exists probability space such that all observables a, b_1, b_2 are represented by random variables $j(a), j(b_1), j(b_2)$. There is well-defined joint probability distribution $p^{ab_1 b_2}(\alpha, \beta_1, \beta_2)$. The condition of consistency of distributions (5.67) implies that

$$p_{C_1}^{(ab_1)}(\alpha, \beta_1) = \sum_{\beta_2} p^{ab_1 b_2}(\alpha, \beta_1, \beta_2), \tag{5.68}$$

$$p_{C_2}^{(ab_2)}(\alpha, \beta_2) = \sum_{\beta_1} p^{ab_1 b_2}(\alpha, \beta_1, \beta_2) \tag{5.69}$$

and

$$p_{C_1}^{(a)}(\alpha) = \sum_{\beta_1, \beta_2} p^{ab_1 b_2}(\alpha, \beta_1, \beta_2). \tag{5.70}$$

Thus we obtain the condition of marginal consistency

$$\sum_{\beta_1} p_{C_1}^{(ab_1)}(\alpha, \beta_1) = \sum_{\beta_1, \beta_2} p^{ab_1 b_2}(\alpha, \beta_1, \beta_2) = \sum_{\beta_2} p_{C_2}^{(ab_2)}(\alpha, \beta_2). \tag{5.71}$$

This is the trivial consequence of Kolmogorovness.

We turn again to Example 2. If contexts $C_{\theta, \gamma}$ were probabilistically compatible, then the condition of marginal consistency would have the form:

$$\sum_{\beta} p_{C_{\theta \gamma_1}}^{a_\theta b_{\gamma_1}}(\alpha, \beta) = \sum_{\beta} p_{C_{\theta \gamma_2}}^{a_\theta, b_{\gamma_2}}(\alpha, \beta). \tag{5.72}$$

In the same way

$$\sum_{\alpha} p_{C_{\theta_1 \gamma}}^{a_{\theta_1}, b_\gamma}(\alpha, \beta) = \sum_{\alpha} p_{C_{\theta_2 \gamma}}^{a_{\theta_2}, b_\gamma}(\alpha, \beta). \tag{5.73}$$

We know that quantum probabilities for the EPR-Bohm-Bell experiment satisfy these conditions. However, there exist triples of contexts $C_{\theta\gamma}, C_{\gamma\phi}, C_{\theta\phi}$ which violate Bell's inequality, e.g., in Wigner's form, see (1.14), section 1.3.2. Therefore these contexts are not probabilistically compatible. It is impossible to construct Kolmogorov probability space matching Definition 2 (otherwise we were able to derive the Bell inequality).

It may be better to speak about quadruples of contexts

$$C_{\theta_1\gamma_1}, C_{\theta_2\gamma_1}, C_{\theta_1\gamma_2}, C_{\theta_2\gamma_2}$$

which violate CHSH-inequality (Chapter 8). The latter can be tested experimentally. However, here one should be careful by identifying the theoretical and experimental violations of the Bell inequality, since experimental setups generate so called loopholes. Their presence change the structure of reasoning completely.

5.5.3.2　*Conditional compatibility*

The notion of unconditional probabilistic compatibility (Definition 2) formalizes the unified description of a few contexts by using classical Kolmogorov probability theory. Discussions about the possibility of such a description may arise in experiments involving a few experimental contexts $C_1, ..., C_m$. Here one investigates whether the data collected in some measurements $(C_i, b_i), b_i \in O_{C_i}$ can be described in the noncontextual framework, i.e., whether it is possible to ignore context-labeling of observables. The notion formalized in Definition 2 is a step in this direction. It matches the standard hidden variable studies in quantum foundations and quantum information theory. The heuristic behind hidden variables and "Definition 2 compatibility" is that physicists proceeded similarly in classical statistical physics and statistical thermodynamics. In such an approach the contextual nature of measurements is ignored; contextual probabilities corresponding to various contexts are treated as *marginal probabilities*, see (5.68) - (5.70) of some "unifying probability distribution." However, if one proceeds another way around, starting with the contextual probabilistic description, then the following natural question arises:

"*How often are contexts $C_i, i = 1, 2, ..., m$, selected in a multi-context experiment?*"

Surprisingly this question is totally ignored in quantum experimenting, in such basic experiments as, e.g., tests on violation of the Bell inequality.

The straightforward explanation of this ignorance is that the formalism of QM by itself does not contain the description of context selection in multi-context experiments. Of course, experimenters are conscious about selection of contexts, e.g., $C_{\theta_i \gamma_j}$ in the Bell test, and even randomize this process, i.e., they are even aware that there are probabilities of selections, $p_1, ..., p_m$. However, they are then able to suppress this information and it is absent in the corresponding theoretical description of multi-context experiments.

We start with the contextual probabilistic model and in this approach the probabilities of context selection have to play the key role. Roughly speaking from our viewpoint it is meaningless to try to proceed with the "Definition 2 type compatibility", since here the aforementioned key element is missing.

Consider a contextual probability model $\mathcal{P}_C = (\mathcal{C}, \mathcal{O}_C, \pi)$. We are now interested in (im)possibility to represent some family of contexts of this model with the aid of single Kolmogorov probability space by taking into account the procedure of context-selection.

Definition 3. *A family of contexts $\tilde{C} = \{C_i\}$ belonging C is called (conditionally) probabilistically compatible with probabilities of conditioning $\{p_i\}$ if there exists Kolmogorov probability space $\mathcal{P} = (\Omega, \mathcal{F}, P)$ such that all observables with respect to these contexts can be represented in $R(\mathcal{P})$, see (5.65), and there exists a "context selection random variable" η taking values i (indexing contexts) such that for each C_i and $b \in O_{C_i}$,*

$$p_{C_i}^b(\beta) = P(j(b) = \beta | \eta = i). \tag{5.74}$$

Here the probability in the right-hand side of (5.74) is the classical conditional probability given by the Bayes formula,

$$P(j(b) = \beta | \eta = i) = \frac{P(\omega \in \Omega : j(b)(\omega) = \beta, \eta(\omega) = i)}{P(\omega \in \Omega : \eta(\omega) = i)}. \tag{5.75}$$

The probability in denominator equals to the context selection probability p_i. Thus, in the case of conditional probabilistic compatibility we have:

$$p_{C_i}^b(\beta) = \frac{1}{p_i} P(j(b) = \beta, \eta = i). \tag{5.76}$$

For example, for the Bell test (for the CHSH inequality) conditional probabilistic compatibility would imply that there exists Kolmogorov probability space where all observables a_θ, b_γ are represented as random variables and there exists a context selection random variable $\eta = (\eta_a, \eta_b), \eta_a, \eta_b = 1, 2$ such that

$$p_{C_{\theta \gamma_1}}^{a_{\theta_i} b_{\gamma_j}}(\alpha, \beta) = P(a_{\theta_i} = \alpha, b_{\gamma_j} = \beta | \eta_a = i, \eta_b = j) \tag{5.77}$$

or

$$p_{C_{\theta\gamma_1}}^{a_{\theta_i}b_{\gamma_j}}(\alpha,\beta) = \frac{P(\omega \in \Omega : a_{\theta_i}(\omega) = \alpha, b_{\gamma_j}(\omega) = \beta, \eta_a(\omega) = i, \eta_b(\omega) = j)}{P(\omega \in \Omega : \eta_a(\omega) = i, \eta_b(\omega) = j)}.$$

(5.78)

5.5.3.3   *On inter-relation of classical and quantum probability*

It is possible to prove that in all natural situations *contexts are always conditionally probabilistically compatible.* In Chapter 8 we shall present the construction of corresponing probability space for the Bell test [168], and the two-slit experiment (see paper [170]). *Thus any multi-context experiment admits the classical probabilistic (Kolmogorov measure theoretic) description, but experimental probabilities have to be treated as conditional with respect to selections of experimental contexts.*

From the viewpoint of classical probability theory the calculus of quantum probabilities is a special (in fact, very special) calculus of classical conditional probabilities. Of course, the calculus of conditional (classical) probabilities differs crucially from the calculus of so to say absolute probabilities. In particular, the presence of the nontrivial Bayesian denominator in (5.76) can lead to increasing of probabilities and correlations. For example, see Chapter 8, where we present the viewpoint that quantum correlations for entangled states are "too strong" comparing with classical correlations, precisely because they are (classical) conditional correlation, not absolute correlations. Of course, they are too strong for absolute correlations, but their magnitudes match well with magnitudes of classical conditional correlations.

Of course, one may question the whole idea of the conditional (in)compatibility and, hence, classical representation of quantum probabilities as conditional (Bayseian) probabilities. If one uses the notion of unconditional compatibility, noncontextuality, as the basis to characterize distinguishable features of quantum probabilities and correlations, then these features would seem to be mystical. However, in the consistent contextual framework not unconditional, but conditional (in)compatibility matches the real experimental situation.

5.6   Quantum Formula of Total Probability

In section 5.1 by using the QM formalism for the two-slit experiment (and following Feynman) we derived generalization of classical FTP, FTP with the additional interference term, (5.2). In the following sections we showed that the same formula, see (5.23), can be derived in the abstract contextual framework, i.e., without exploring the mathematical formalism of QM. Unfortunately, we were able to succeed only in the case of dichotomous observables.

Now by exploring QM we derive FTP with interference terms for arbitrary discrete quantum observables.

5.6.1 *Interference of von Neumann-Lüders observables*

We start with derivation of quantum FTP for quantum instruments a, b of the von Neumann-Lüders type, i.e., the observables are given by Hermitian operators A, B with the spectral decompositions $A = \sum_k \alpha_k P_{\alpha_k}, B = \sum_j \beta_j Q_{\beta_j}$. Here quantum operations (state transformations) are given by the projectors corresponding to eigenvalues (we recall that we consider the finite dimensional case). We also recall that, for each Hermitian operator the sum of projectors corresponding to its eigenvalues equals to the unit operator I :

$$\sum_k P_{\alpha_k} = I, \sum_j Q_{\beta_j} = I. \tag{5.79}$$

Let ρ be an arbitrary quantum state (density operator). Derivation of quantum FTP is similar to the case of a pure state ψ [141]. We consider the outputs of the a-instrument, the states

$$\rho_{\alpha_k} = \frac{P_{\alpha_k} \rho P_{\alpha_k}}{\mathrm{Tr} P_{\alpha_k} \rho P_{\alpha_k}}$$

and outputs of the b-instrument corresponding to the inputs ρ_{α_k},

$$\rho_{\beta_j|\alpha_k} = \frac{Q_{\beta_j} \rho_{\alpha_k} Q_{\beta_j}}{\mathrm{Tr} Q_{\beta_j} \rho_{\alpha_k} Q_{\beta_j}} = \frac{Q_{\beta_j} P_{\alpha_k} \rho P_{\alpha_k} Q_{\beta_j}}{\mathrm{Tr} Q_{\beta_j} P_{\alpha_k} \rho P_{\alpha_k} Q_{\beta_j}}. \tag{5.80}$$

These states are produced in the following way. First a-measurement is performed and the result $a = \alpha_k$ is detected. The feedback of this measurement transfers the initial state ρ into the state ρ_{α_k}. Then for this state b-measurement is performed and the result $b = \beta_j$ is detected. The feedback of the latter measurement transfers the state ρ_{α_k} into the state $\rho_{\beta_j|\alpha_k}$.

To shorten notations, we shall often omit the state index for probabilities.

By using the equality (5.79) and the Born rule for the b-observable, we obtain:

$$p(b = \beta) = \mathrm{Tr} Q_{\beta_j} \rho = \sum_k \sum_m \mathrm{Tr} Q_{\beta_j} P_{\alpha_k} \rho P_{\alpha_m}. \tag{5.81}$$

This probability can be represented as the sum of the classical (diagonal) and nonclassical (non-diagonal) counterparts:

$$p(b = \beta_j) = p_{\text{diag}}(b = \beta_j) + p_{\text{ndiag}}(b = \beta_j). \tag{5.82}$$

We start with the classical part of the sum given by the right-hand side of (5.81):

$$p_{\text{diag}}(b = \beta_j) = \sum_k \frac{\text{Tr}Q_{\beta_j}P_{\alpha_k}\rho P_{\alpha_k}}{\text{Tr}P_{\alpha_k}\rho P_{\alpha_k}} \text{Tr}P_{\alpha_k}\rho P_{\alpha_k} = \sum_k p(b = \beta_j | a = \alpha_k)p(a = \alpha_k)$$

(if the quantities in the denominator differ from zero).

Now we rewrite the nonclassical part of the sum as

$$p_{\text{ndiag}}(b = \beta_j) = \sum_{k \neq m} \text{Tr}Q_{\beta_j}P_{\alpha_k}\rho P_{\alpha_m} = \sum_{k \neq m} \text{Tr}Q_{\beta_j}P_{\alpha_k}\rho P_{\alpha_m}Q_{\beta_j},$$

here we used the cyclic property of the trace and idempotence of projectors.

It is convenient to introduce in the space of bounded operators the following form (cf. (4.65), section 4.12):

$$\langle C|D\rangle_\rho = \text{Tr}C^\star \rho D. \tag{5.83}$$

It has the following properties:

- skew-symmetry: $\langle C|D\rangle_\rho = \overline{\langle D|C\rangle_\rho}$ (this is a consequence of the equality $\text{Tr}K^\star = \overline{\text{Tr}K}$);
- positive-semidefiniteness: $\langle C|C\rangle_\rho \geq 0$.

Hence, this is "practically a scalar product", but it can be degenerate, i.e., $\langle C|C\rangle_\rho = 0$ does not imply that $C = 0$.

With this notation we write

$$p_{\text{ndiag}}(b = \beta_j) = \sum_{k \neq m} \langle P_{\alpha_k}Q_{\beta_j}|P_{\alpha_m}Q_{\beta_j}\rangle_\rho$$

$$= 2\sum_{k < m} |\langle P_{\alpha_k}Q_{\beta_j}|P_{\alpha_m}Q_{\beta_j}\rangle_\rho| \cos \gamma_{j:km},$$

where the phase $\gamma_{j:km}$ equals to the argument of the complex number $\langle P_{\alpha_k}Q_{\beta_j}|P_{\alpha_m}Q_{\beta_j}\rangle_\rho$. Now by using the *Cauchy-Bunyakovsky-Schwartz inequality* (which holds even for degenerate positive-semidefinite skew symmetric forms) we obtain that

$$k_{j;km} = \frac{|\langle P_{\alpha_k}Q_{\beta_j}|P_{\alpha_m}Q_{\beta_j}\rangle_\rho|}{\sqrt{\langle P_{\alpha_k}Q_{\beta_j}|P_{\alpha_k}Q_{\beta_j}\rangle_\rho \langle P_{\alpha_m}Q_{\beta_j}|P_{\alpha_m}Q_{\beta_j}\rangle_\rho}} \leq 1 \tag{5.84}$$

(if the quantities in denominator differ from zero). The expression $k_{j;km} \cos \gamma_{j;km}$ can be represented as cosine of some angle:

$$k_{j;km} \cos \gamma_{j;km} = \cos \phi_{j;km}.$$

Hence, with this notation we obtain the quantum FTP:

$$p(b = \beta) = \sum_k p(b = \beta | a = \alpha_k) p(a = \alpha_k) \tag{5.85}$$

$$+ 2 \sum_{k<m} \cos \phi_{j;k,m} \sqrt{p(b = \beta | a = \alpha_k) p(a = \alpha_k) p(b = \beta | \alpha_m) p(a = \alpha_m)}.$$

Thus, for observables of the von Neumann-Lüders type, the classical FTP is simply additively deformed with the additional interference term which depends nonlinearly on probabilities.

To compare with the wave function calculations for the two-slit experiment and with contextual probability modification of FTP (section 5.4.2), we specify the above quantum FTP to the case of two dichotomous observables of the von Neumann-Lüders type, $a = \alpha_1, \alpha_2, b = \beta_1, \beta_2$. To shorter notation and to match the notations of section 5.4.2, we denoted the quantum conditional probability (section 4.3) $p(b = \beta | b = \alpha)$ by the symbol $p_{\beta | \alpha}$ and set $p^a(\alpha) \equiv p(a = \alpha), p^b(\beta) \equiv p(b = \beta)$. We have

$$p^b(\beta) = p_{\beta_j | \alpha_1} p^a(\alpha_1) + p_{\beta_j | \alpha_2} p^a(\alpha_2)$$

$$+ 2 \cos \theta_j \sqrt{p_{\beta_j | \alpha_1} p^a(\alpha_1) p_{\beta_j | \alpha_2} p^a(\alpha_2)}. \tag{5.86}$$

Here we denoted the "phase-angles" by θ_j, because it can be checked that in this case the k-coefficients, see (5.84), are equal to 1. We can express the interference term as

$$\lambda_j \equiv \cos \theta_j = \frac{p^b(\beta) - [p_{\beta_j | \alpha_1} p^a(\alpha_1) + p_{\beta_j | \alpha_2} p^a(\alpha_2)]}{2\sqrt{p_{\beta_j | \alpha_1} p^a(\alpha_1) p_{\beta_j | \alpha_2} p^a(\alpha_2)}}. \tag{5.87}$$

The formulas (5.86) and (5.87) coincide with the corresponding formulas of section 5.4.2 devoted to the *quantum-like representation* of contextual probabilistic data. In the quantum approach context C is encoded by the quantum state used for measurements, by the density operator ρ. To index contextuality of probabilities we can write $p^a(\alpha) \equiv p^a(\alpha | \rho), p^b(\beta) \equiv p^b(\beta | \rho)$.

The derivation of FTP with the interference term, "quantum FTP", illustrates an important feature of the inverse Born problem (IBP), see section 5.4.1. We can see that probabilistic data determine uniquely the quantum state only under restriction of its purity. QLRA produces pure states. Now suppose that, as in this section, we started with a state given by the density operator ρ. Then starting with (5.86), (5.87) QLRA produces the pure state $\psi \equiv \psi_\rho$ (here context $C = \rho$). However, if ρ was not pure, then by solving IBP we do not reconstruct the original input state ρ. We may say that IBP is the ill-posed problem and that QLRA is one of its possible regularizations.

In this situation one may question the usefulness of QLRA: "the real state of a system" is mixed, but it is represented as a pure state! Here it is important to remind the quantum-like ideology. We construct complex state space starting with data. To obtain appropriate state space, it is natural to start with data representing the basic states of a dynamical system generating these data. In the quantum-like representation these states appear (as outputs of QLRA) as pure states. Then "mixed states" appear as probabilistic mixtures of the basic states and they are represented by density operators.

5.6.2 *Interference of positive operator valued measures*

The situation is more complicated for generalized observables given by POVMs. Here even the diagonal part of quantum FTP does not coincide with classical FTP.

Consider two atomic instruments. The observables are given by POVMs

$$a = (M_j^a = A_j^\star A_j), b = (M_j^b = B_j^\star B_j)$$

and the quantum operations realized with the aid of the operators A_j and B_j :

$$\rho \to \frac{A_j \rho A_j^\star}{\operatorname{Tr} A_j \rho A_j^\star}, \rho \to \frac{B_j \rho B_j^\star}{\operatorname{Tr} B_j \rho B_j^\star}.$$

We also recall the normalization conditions

$$\sum_k M_k^a = \sum_k A_k^\star A_k = I, \sum_k M_k^b = \sum_k B_k^\star B_k = I.$$

By using the first of them we obtain the following representation of Born's rule:

$$p(b = \beta_j) = \operatorname{Tr} M_j^b \rho = \sum_k \sum_m \operatorname{Tr} M_j^b A_k^\star A_k \rho A_m^\star A_m. \qquad (5.88)$$

Consider first the diagonal term

$$p_{\text{diag}}(b = \beta_j) = \sum_k \operatorname{Tr} M_j^b A_k^\star A_k \rho A_k^\star A_k$$

$$= \sum_k \left(\frac{\mathrm{Tr} M_j^b A_k^\star A_k \rho A_k^\star A_k}{\mathrm{Tr} A_k^\star A_k \rho A_k^\star A_k} \right) \left(\frac{\mathrm{Tr} A_k^\star A_k \rho A_k^\star A_k}{\mathrm{Tr} A_k \rho A_k^\star} \right) \mathrm{Tr} A_k \rho A_k^\star.$$

Consider first $\mathrm{Tr} A_k \rho A_k^\star = \mathrm{Tr} A_k^\star A_k \rho = p(a = \alpha_k)$. Then $\mathrm{Tr} A_k^\star A_k \rho A_k^\star A_k = \mathrm{Tr} A_k A_k^\star A_k \rho A_k^\star$.

We now have to introduce a new quantum instrument. It represents the observable which is given by POVM $\bar{a}$ ("conjugate" to a)

$$\bar{a} = (M_j^{\bar{a}} = A_j A_j^\star) \tag{5.89}$$

and the corresponding quantum operation

$$\rho \to \frac{A_j^\star \rho A_j}{\mathrm{Tr} A_j^\star \rho A_j}. \tag{5.90}$$

Thus

$$\frac{\mathrm{Tr} A_k^\star A_k \rho A_k^\star A_k}{\mathrm{Tr} A_k \rho A_k^\star} = \frac{\mathrm{Tr} A_k A_k^\star A_k \rho A_k^\star}{\mathrm{Tr} A_k \rho A_k^\star} = p(\bar{a} = \alpha_k | a = \alpha_k).$$

Finally, we consider the "doubly conditioned" probability

$$\frac{\mathrm{Tr} M_j^b A_k^\star A_k \rho A_k^\star A_k}{\mathrm{Tr} A_k^\star A_k \rho A_k^\star A_k} = p(b = \beta_j | a = \alpha_k, \bar{a} = \alpha_k).$$

Here $p(b = \beta_j | a = \alpha_k, \bar{a} = \alpha_k) \equiv p(b = \beta_j | \text{first } a = \alpha_k \text{ then } \bar{a} = \alpha_k)$. Hence,

$$p_{\mathrm{diag}}(b = \beta_j) = \sum_k p(b = \beta_j | a = \alpha_k, \bar{a} = \alpha_k) p(\bar{a} = \alpha_k | a = \alpha_k) p(a = \alpha_k).$$

Now we consider the non-diagonal term:

$$p_{\mathrm{ndiag}}(b = \beta_j) = \sum_{k \neq m} \mathrm{Tr} B_j^\star B_j A_k^\star A_k \rho A_m^\star A_m = \sum_{k \neq m} \mathrm{Tr} B_j A_k^\star A_k \rho A_m^\star A_m B_j^\star$$

$$= \sum_{k \neq m} \mathrm{Tr} (A_k^\star A_k B_j^\star)^\star \rho A_m^\star A_m B_j^\star = \sum_{k \neq m} \langle A_k^\star A_k B_j^\star | A_m^\star A_m B_j^\star \rangle_\rho$$

$$= \sum_{k < m} |\langle A_k^\star A_k B_j^\star | A_m^\star A_m B_j^\star \rangle_\rho| \cos \gamma_{j;km},$$

where the phase $\gamma_{j;km}$ equals to the argument of the complex number $\langle A_k^\star A_k B_j^\star | A_m^\star A_m B_j^\star \rangle_\rho$. by using the Cauchy-Bunyakovsky-Schwartz inequality we obtain that

$$k_{j;km} = \frac{|\langle A_k^\star A_k B_j^\star | A_m^\star A_m B_j^\star \rangle_\rho|}{\sqrt{\langle A_k^\star A_k B_j^\star | A_k^\star A_k B_j^\star \rangle_\rho \langle A_m^\star A_m B_j^\star | A_m^\star A_m B_j^\star \rangle_\rho}} \leq 1.$$

The expression $k_{j;km} \cos \gamma_{j;km}$ can be represented as cosine of some angle $\phi_{j;km}$. Hence, we come to following quantum FTP:

$$p(b = \beta_j) = \sum_k p(b = \beta_j | a = \alpha_k, \bar{a} = \alpha_k) p(\bar{a} = \alpha_k | a = \alpha_k) p(a = \alpha_k)$$

$$+2 \sum_{k<m} \cos \phi_{j;km}$$

$$\times \sqrt{\prod_{i=k,m} p(b = \beta_j | a = \alpha_i, \bar{a} = \alpha_i) p(\bar{a} = \alpha_i | a = \alpha_i) p(a = \alpha_i)}. \qquad (5.91)$$

The main difference from the quantum FTP for observables of the von Neumann-Lüders type and from the classical FTP is not in the interference term, but in the diagonal, "classical-like term". Here the reference measurement is based on a single POVM a, but also on its "conjugate POVM" $\bar{a}$.

We remark that all previous derivations can be performed even in the infinite-dimensional case. Here ρ is a trace class operator. Therefore the form (5.83) is well defined for pairs of bounded operators.

One may be surprised that transition from von Neumann-Lüders to arbitrary (atomic) instruments changes the form of quantum FTP so much. However, this difference is easily explainable: this is a consequence non-repeatability. In the mathematical terms this is a consequence of non-idempotency of operators A_{α_j} and B_{β_j} representing quantum operations of atomic instruments. This implies double conditioning as well as conditional probabilities of the form $p(\bar{a} = \alpha_i | a = \alpha_i)$. The main aesthetic element of previous considerations is the appearance of quantum instrument (5.89), (5.90). Theory would be esthetically more attractive if, instead of the probabilities $p(\bar{a} = \alpha_i | a = \alpha_i), p(b = \beta_j | a = \alpha_i, \bar{a} = \alpha_i)$, we were able to proceed with the probabilities $p(a = \alpha_i | a = \alpha_i), p(b = \beta_j | a = \alpha_i, a = \alpha_i) \equiv p(b = \beta_j | \text{first } a = \alpha_k \text{ then again } a = \alpha_k)$.

Now consider the case of atomic instruments with dichotomous generalized observables (POVMs)

$$a = (M_1^a = A_1^\star A_1, M_2^a = A_2^\star A_2), M_1^a + M_2^a = I;$$

$$b = (M_1^b = B_1^\star B_1, M_2^b = B_2^\star B_2), M_1^b + M_2^b = I.$$

Here besides quantum FTP (5.91) taking into account non-reproducibility, we can write "straightforward modification" of classical FTP, cf. with (5.86),

$$p^b(\beta) = p_{\beta_j | \alpha_1} p^a(\alpha_1) + p_{\beta | \alpha_2} p^a(\alpha_2) \qquad (5.92)$$

$$+2\lambda_j \sqrt{p_{\beta|\alpha_1}p^a(\alpha_1)p_{\beta_j|\alpha_2}p^a(\alpha_2)},$$

where

$$\lambda_j = \frac{p^b(\beta) - [p_{\beta_j|\alpha_1}p^a(\alpha_1) + p_{\beta|\alpha_2}p^a(\alpha_2)]}{2\sqrt{p_{\beta|\alpha_1}p^a(\alpha_1)p_{\beta_j|\alpha_2}p^a(\alpha_2)}}. \tag{5.93}$$

It can be shown that, for generalized observables (POVMs), this interference terms can exceed 1 - in contrast to observables of the von Neumann-Lüders type, see (5.87). As we have seen, this can also happen in general contextual probabilistic model, see section 5.3. One might hope to model operationally "too strong contextual interference" with the aid of theory of atomic instruments. However, the situation is more complicated. Our contextual probabilistic model was built under the assumption of repeatability of observables. However, as was already mentioned, in the finite-dimensional case repeatable atomic instruments are of the von Neumann-Lüders type. However, for the latter the interference coefficients cannot exceed 1. In short, it is impossible to model FTP for repeatable observables with the interference term exceeding 1 with the aid of standard QM (at least for discrete observables taking finite number of values). Thus one has to search for generalizations of QM. One of possible quantum-like models solving this problem, repeatability combined with "too strong interference", is the model of *hyperbolic* QM [161], section 5.3.4.

We remark that, for some statistical data collected in cognitive psychology, the interference terms exceed 1 [163]. At the same time observables are repeatable. It seems that such data cannot be described either by classical or quantum probability, see Chapter 9 on quantum-like models of cognition and psychology. It may be that hyperbolic QM is really an adequate model for such phenomena [163].

Chapter 6

Interpretations of Quantum Mechanics and Probability

The modern situation in QM is characterized by the diversity of its interpretations. Moreover, as was remarked in my book [167] (and section 5.2.2) each basic interpretation has a variety of individual flavors. For example, suppose that somebody claims to be an adherent of the Copenhagen interpretation of QM. When you ask for details, most probably you will meet with a very specific view on what this interpretation is about. I have the impression that there are so many versions of the Copenhagen interpretation of QM as there are people knowing about it. The same can be said about, e.g., the statistical or Bohmian interpretations of QM.

6.1 Classification of Interpretations

6.1.1 *Realism and reality*

We shall try to classify the interpretations of QM. There are many ways to do this. For example, we can split the set of interpretations into two classes: realist and non-realist interpretations. The notion of realism in its relation to quantum foundations has some degree of ambiguity. I like the "mathematical model" viewpoint on realism, see Plotnitsky and Khrennikov [223]:

- **R**: *Realist interpretation.* The ultimate nature of quantum systems and processes is assumed to be representable by means of the mathematical formalism of quantum mechanics or by some subquantum mathematical model.
- **NR**: *Non-realist interpretation.* Negation of the realist interpretation.

In particular, **NR** means that, as Bohr emphasized, QM does not describe quantum systems and processes as "they are", but it describes (and predicts) the outputs of measurements performed on quantum systems (and measurement apparatuses are treated classically), see Bohr [39], [40]:

"This crucial point, which was to become a main theme of the discussions reported in the following, implies the impossibility of any sharp separation between the behavior of atomic objects and the interaction with the measuring instruments which serve to define the conditions under which the phenomena appear. In fact, the individuality of the typical quantum effects finds its proper expression in the circumstance that any attempt of subdividing the phenomena will demand a change in the experimental arrangement introducing new possibilities of interaction between objects and measuring instruments which in principle cannot be controlled. Consequently, evidence obtained under different experimental conditions cannot be comprehended within a single picture, but must be regarded as complementary in the sense that only the totality of the phenomena exhausts the possible information about the objects."

It is interesting that, for Bohr, non-realist picture of quantum phenomena is an evident consequence of his principle of complementarity which in turn is a consequence of *Heisenberg's uncertainty relation*. For example, one cannot treat the position and momentum of an electron in the realist way, as its objective properties. By using philosophic terminology we say that QM does not provide the *ontic description* of phenomena - as they are, without relation to observations and observers. QM is an *epistemic description*, see Primas [224] and Atmanspacher et al. [20], [21].

We have discussed negation of only the first part of the formulation of the realist interpretation. Negation of the second part declares the impossibility of construction of any subquantum mathematical model which would represent "the ultimate nature of quantum systems and processes". This (an essentially stronger) form of nonrealism is common for the modern interpretations of QM in the spirit of Copenhagen.[1] One of its forms is negation of the possibility to construct a complete model with hidden variables. The latter is defined as follows (the definition of a model with hidden variables):

(1) The physical properties of a systmem S are completely specified by a

---

[1] This expression "in the spirit of Copenhagen" was invented by A. Plotnitsky in his talks during the Växjö series of conferences on quantum foundations. It is very useful to label a variety of views and interpretations related to the Copenhagen school of thought (Bohr, Heisenberg, Pauli, Dirac, von Neumann, Landau, Fock, ...) including neo-Copenhagenists (Mermin, Zeilinger, Fuchs, Brukner, Gill, Schack, Bengtsson, Larsson,...).

variable λ.

(2) For any measurement M, the probability of its outcome k is determined by λ.

In its extreme form the non-realist interpretation denies the possibility to create any mathematical model of quantum phenomena as independent from the interventions of measurement devices. We remark that *such models are not restricted to the models with hidden variables*, i.e., straightforwardly coupling quantum world with the results of measurements, see section 6.1.4 for discussion.

We also remark that the terminology "subquantum model" has to be used with caution. A model is "subquantum" with respect to the mathematical model of QM, i.e., "subquantum" means a deeper description than provided by the standard quantum formalism. However, in its physical essence a subquantum model is "really quantum", i.e., it should provide (if it were finally created) a realist description of quantum systems and processes, the ontic description of quantum reality - as it is, without influence of our macroscopic classical measurement devices.

Realist and non-realist interpretations have numerous personal flavors:

- **R**: Planck, Einstein, De Broglie, Schrödinger, von Neumann, Bohm, Blohintzev, Ballentine, Rauch, Lahti and Mittelstandt, Vaidman, Accardi, Gisin, Khrennikov, Kupczynski, Hess and Philipp, Nieuwenhuizen, De Raedt, ...;
- **NR**: Bohr, Heisenberg, Pauli, Fock, von Neumann, Dirac, Landau, Mermin, Belavkin, Holevo, Zeilinger, Brukner, Weihs, Kofler, Kwiat, Mermin, Caves, Fuchs, Schack, Gill, D' Ariano, Bengtsson, Plotnitsky, Larsson, ...

P.S. The presence of von Neumann in both lists is not a misprint, see section 6.3.

In discussions on quantum foundations one has to sharply distinguish reality and realism, see the article [223] with the intriguing title "Reality without realism". One may think that the members of the Copenhagen circle were idealists. This is not true. Of course, Bohr, Heisenberg, Pauli, Landau were completely sure that, e.g., atoms and electrons exist. They simply questioned the possibility of creation of a realist model in the sense of **R**, i.e., a model which would not take into account the contribution of measurement devices into values of quantum observables. Moreover, for Copenhagenists, the results of measurements were objective.

It is also important to remark that the **R**-definition cannot pretend to be the only possible definition of a realist interpretation. QBists (section 6.7) consider their interpretation of QM as a realist one. However, they assign totally different meaning to realism, as *objective indeterminism*, see James [120], [121].

6.1.2 *Epistemic and ontic description*

The above discussion of reality and realism is closely related to discussions, favored by philosophers of science, on *ontic* and *epistemic descriptions*. The most clear and complete presentation of the interplay between ontic and epistemic levels of description in quantum and classical physics can be found in works Primas [224] and H. Atmanspacher et al. [20], [21]. We start with citation from the paper of Atmanspacher et al. [20], p. 53 (see also Primas [224] for more details):

"Ontic states describe all properties of a physical system exhaustively. ("Exhaustive" in this context means that an ontic state is "precisely the way it is", without any reference to epistemic knowledge or ignorance.) Ontic states are the referents of individual descriptions, the properties of the system are treated as intrinsic properties. Their temporal evolution (dynamics) is reversible and follows universal, deterministic laws. As a rule, ontic states in this sense are empirically inaccessible. Epistemic states describe our (usually non-exhaustive) knowledge of the properties of a physical system, i.e., based on a finite partition of the relevant phase space. The referents of statistical descriptions are epistemic states, the properties of the system are treated as contextual properties. Their temporal evolution (dynamics) typically follows phenomenological, irreversible laws. Epistemic states are, at least in principle, empirically accessible."

In principle, one can treat subquantum models and, in particular, models with hidden variables, as providing the ontic description of microphenomena and the quantum model provides the epistemic description. The separate treatment of the ontic and epistemic descriptions of reality played an important role in making foundational discussions on QM less ambiguous. Atmanspacher correctly pointed out that the inability to distinguish ontic from epistemic was the main root of total misunderstanding and, as the result, total miscommunication between Einstein and Bohr in their debate on the EPR- experiment. Bohr viewed QM as an epistemic model and Einstein as an ontic model. And they both were right, QM is complete from the epistemic viewpoint and not from the ontic viewpoint. Personally I liked the ontic-epistemic viewpoint to quantum foundations elaborated by Primas [224] and Atmanspacher et al. [20], [21]. Moreover, Atmanspacher et al. [20] presented the clear and explicit approach to the ontic-epistemic treatment of classical physics: the ontic description is the deterministic Newtonian model and the epistemic description is provided by statistical physics. Thus it seems that this approach works well in the whole domain of physics. I found it also useful in cognition studies [163].

However, one thing in this ontic-epistemic approach was always disturbing me. This is the assignment of some special value to an ontic state: that an ontic state is "precisely the way it is". It seems that Primas [224] and Atmanspacher et al. [20], [21] and others following the ontic-epistemic approach assign a too high value to the ontic states and the ontic level of description.

My own ideology of scientific studies can be called the *mathematical modeling approach*. The only way to make research is by constructing mathematical models. In this approach there are no ontic states describing all properties of a physical system exhaustively, because we even cannot speak about properties without first introducing a mathematical concept of "properties". And no mathematical model can represent properties of a system exhaustively. In this approach we define "real" only what complies perfectly with our model. There are only mathematical symbols representing our knowledge about reality. In some sense all descriptions, since they are meaningful only as mathematical formalisms, are epistemic.

Consider, for example, classical electrodynamics. Here the basic physical entity is the *electromagnetic field*, its state represented by components of the electric and magnetic fields. Do these components represent some intrinsic properties of the field? I am not sure. At the very beginning when the field was considered as an oscillating wave in *aether* it was still possible to assign to it some degree of "onticity" - existence as it is, as a wave of matter oscillations. However, nowadays when it is forbidden to appeal to aether, the field is just the vector of mathematical symbols $\phi(x) = (E_j(x), B_j(x))$. The Maxwell dynamics is, of course, deterministic, but it represents just our knowledge about behavior of these mathematical symbols. By considering random fields we depart from determinism, but knowledge gained from the random field model (which is probabilistic) is knowledge about the same mathematical symbols. Thus neither deterministic nor random field models can be treated as describing reality as it is, when nobody looks at it.

Thus, may be the whole ontic-epistemic approach [224], [20], [21] is misleading? May be it is better simply to speak about deterministic (or more generally causal) model $\mathcal{M}_d$ and statistical model $\mathcal{M}_{st}$. The first class of models can be treated as realist models. What is about the second class? If a statistical model $\mathcal{M}_{st}$ can be constructed on the basis of a deterministic model $\mathcal{M}_d$ as the result of uncertainty in initial and boundary conditions, then we just say that $\mathcal{M}_{st}$ is generated by $\mathcal{M}_d$, see [167] (and section 5.2.2). We cannot say anything more, both $\mathcal{M}_d$ and $\mathcal{M}_{st}$ are equally just our mathematical representations of reality.

It is important to remark that determinism of $\mathcal{M}_d$ is meaningful only internally in this model. It may happen that the model $\mathcal{M}_d$ by itself can be represented as a statistical model with respect to another deterministic model $\mathcal{M}'_d$ by assigning to the states of $\mathcal{M}_d$ statistical meaning with respect to the states of $\mathcal{M}'_d$. Then $\mathcal{M}_d$ can be considered as $\mathcal{M}'_{st}$ and so on. There is no bottom level, subquantum models are statistical models for sub-subquantum, then sub-sub-subquantum and so on...

6.1.3 *Individual and statistical interpretations*

We can also split the set of interpretations in other two classes based on the individual and ensemble interpretations of the quantum state (wave function):

- **I**: *Individual interpretation.* A quantum state represents features of the individual quantum system.
- **E**: *Ensemble interpretation.* A quantum state represents statistical features of an ensemble of identical (vs identically prepared) quantum systems.

In **I**, "features" can be treated in various ways, for example, physical features (Dirac, von Neumann, early Bohr), or knowledge (Heisenberg, Zeilinger and Brukner), or subjective experience (e.g., Mermin, Caves, Fuchs, Schack). Personalization of the views leads to the following lists:

- **I**: Schrödinger, Bohr, Dirac, von Neumann, Lahti, Mittelstandt, Zeilinger, Brukner, Belavkin, Mermin, Caves, Fuchs, Schack, ...;
- **E**: Planck, Einstein, Bohr, Bohm, Blohintzev, Ballentine, Rauch, Holevo, Khrennikov, Kupczynski, Hess, Philipp, Nieuwenhuizen, De Raedt, Plotnitsky, ...

Note. The presence of Bohr in both lists is not a misprint, see section 6.3. The presence of Schrödinger in the **I**-list must be taken with caution, since he wanted to treat the wave function as a classical physical field. This viewpoint does not match the viewpoints of others in this list!

I was not able to produce a table combining **R/NR** and **I/E** factorizations, because of the lack of complete information about the personal views.

6.1.4 *Subquantum models and models with hidden variables*

Now we make a comment about the notion of realism by comparing it with the notion of hidden variable. Hidden variables in QM were discussed in details by von Neumann. He considered the wave function $\psi = \psi(q_1, ..., q_n)$ of a quantum system, a function of the coordinates $q_1, ..., q_n$. It determines only the probability to find this system in the point with these coordinates, $p(q_1, ..., q_n) = |\psi(q_1, ..., q_n)|^2$. Then he remarked that one of the possible sources for such randomness can be incompleteness of the description of system's state by using the coordinates. One can imagine that there exist some additional parameters determining the state of the system, parameters not covered by the quantum formalism. By completing the state description with these variables we would be able to determine precisely the coordinates of a system and other quantum quantities, e.g., the momentum and the energy. Such parameters are known as *hidden variables.*

It is useful to present the following comparison with classical mechanics. Here the state is determined by the vectors of the coordinates and momenta. Suppose that only the vector of coordinates is known. Then the state description is not complete and this incompleteness is the source of randomness. However, randomness disappears when the state description is completed with the momentum vector.

It is clear that a theory with hidden variables is realistic. However, in accordance with the above definition theories with hidden variables form just a special class of realist theories. For example, consider my subquantum realistic model known as *prequantum classical statistical field theory* (PCSFT) [167] (see section 5.2.2). Here we simply do not have systems. There is a source of a random physical field. However, this field cannot be separated from the "*zero point field*", the random background field. Even the picture of a source emitting pulses of classical signal interacting with detectors is too naive to match PCSFT. There is simply a random field which produces clicks of detectors in random instances of time. Of course, a field source used in the preparation procedure contributes to this random field, but not straightforwardly. At least partially PCSFT matches the dreams of Einstein and Infeld [86] and Schrödinger [228] - [231] for a purely field model of quantum phenomena. We cannot exclude that imagination of future generations will lead to novel and totally unexpected subquantum models.

6.1.5   *Nonlocality*

Another interpretational factor is *locality* which is also defined very generally as the absence of action at the distance [31], see, e.g., [119] for discussion.[2] Personally I think that nonlocality is solely the product of Bohm's thought, see, e.g., Bohm and Hiley [36], which was lifted to the level of the fundamental problem of QM by Bell [29], [31]. It is interesting that even De Broglie whose theory of the *double solution* was used by Bohm as the initial basis for Bohmian mechanics did not consider QM or his own theory as nonlocal [67]. Both Einstein and Bohr, as well as Heisenberg, Pauli, Fock, von Neumann, Dirac, Landau, Blohintzev, considered QM as a local theory. Nowadays Einstein is typically pointed to as the inventor of the problem of nonlocality to QM. However, practically nobody mentions that the issue of nonlocality was characterized as an *absurd alternative* (to incompleteness of QM) [85]. It is also completely forgotten that Bohr in his reply to Einstein [38] argued that from his viewpoint QM is both complete and local. Nowadays the situation with nonlocality of QM can be characterized as one of the forms of cognitive dissonance.

People claim that they follow Bohr's viewpoint on QM and at the same time they are sure that Einstein's reasoning about "an absurd alternative to incompleteness" implies nonlocality of QM. Einstein would never imagine that this absurd alternative would become commonly acceptable and (what is the most amazing) in combination with completeness of QM. For Einstein, if QM were complete there was no need to consider nonlocality at all.

6.2   Interpretations of Probability and Quantum State

In section 1.1 we pointed out that the diversity of interpretations of a quantum state (wave function) is coupled with the diversity of the inter-

---

[2]In [144], see also the book [161], we criticized this viewpoint on locality as quite different from locality defined in the framework of the relativity theory.

pretations of probability. Now we discuss this problem in more detail. We emphasize that by discussing some interpretation of QM one has to clarify the corresponding interpretation of probability given by the Born rule. Unfortunately, this is not the custom in the quantum community. For some interpretations, one can only guess how probability has to be interpreted.

We recall that there are two basic approaches to probability:

- **ST** Statistical.
- **SUB** Subjective (Bayesian).

By the statistical interpretation probability is a characteristics of "a mass phenomenon or a repetitive event, or simply a long sequence of observations" (von Mises, see citation in Chapter 1). In this approach to probability we can speak only about probability of a repeatable event, and not of an individual event. Mathematically statistical probability is represented either as Kolmogorov measure-theoretic probability or as von Mises frequency probability. By the subjective interpretation probability is a measure of individual belief in (non)occurrence of an individual event. It is interesting that mathematically subjective probability can also be presented in measure-theoretic framework of Kolmogorov (with modifications done by de Finetti [69]).

Personalization of the interpretations of probability given by the Born rule leads to the following lists:

- **ST** Planck, Einstein, Bohr, Pauli, Dirac, von Neumann, Bohm, Blohintzev, Schwinger, Ballentine, Rauch, Holevo, Khrennikov, Kupczynski, Hess, Philipp, Nieuwenhuizen, De Raedt, and Plotnitsky
- **SUB** Heisenberg, Mermin, Caves, Fuchs, Schack, Zeilinger, Brukner.

We see that in quantum foundations the subjective probability interpretation is really underrepresented.

In general, one has to take these personal classifications of interpretations of QM with caution; in particular, they are time dependent. Now we consider a variety of interpretations in the spirit of Copenhagen.

6.3 Orthodox Copenhagen Interpretation

We use the term "orthodox" to distinguish this interpretation from other interpretations in "the spirit of Copenhagen." However, many authors call it simply the Copenhagen interpretation.

This is an individual interpretation of QM. A quantum state (wave function) ψ presents the most complete description of the state of an individual quantum system. One can speak about the wave function of, e.g., a single electron.

This interpretation definitely belongs to the class of non-realist interpretations. In this interpretation it is impossible to invent hidden variables providing a realistic basis for QM. For Bohr and Heisenberg, this is the straightforward consequence of the principle of complementarity and uncertainty relation. Von Neumann was not satisfied by the Bohr-Heisenberg "derivation" of nonrealism from quantum complementarity-uncertainty and he proved the first no-go theorem for hidden variables [250].

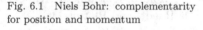

Fig. 6.1 Niels Bohr: complementarity for position and momentum

Fig. 6.2 William James: complementarity for unconscious and conscious

Comment to the picture : One may be surprised to see fathers of QM and modern psychology nearby. Our intention is to underline that Bohr borrowed the principle of complementarity from psychology, where it was first time formalized in the book of James [122].[3] For James, the principle of complementarity was recognition of impossibility to combine in

---

[3]For those who have never heard about William James, we remark that he was one of the greatest psychologists of 19th century. In 20th century he was shadowed by the popularity of Freud (who also took a lot from James). Thus at this picture Bohr is in very good company. It is funny that this picture was one of the reasons of rejection of my (invited) review paper written for "Physica Scripta". The reviewer was angry that I tried to diminish the glory of Bohr as the discoverer of the great *physical principle* of complementarity by comparing it with unscientific writings of psychologists. (Later this paper was published as [125].)

one picture two mental representations - unconscious and conscious. For Bohr, this principle primarily arose as recognition of the impossibility to combine two physical representations - position and momentum. Typically this joint use of the same principle in psychology and physics is treated as just exploration of analogy. However, one may go deeper and speculate that both complementarities are simply two forms of the same complementarity - complementarity in representation of information by humans. It seems that QBists (section 6.7) would support this statement (probably, together with adherents of the information interpretation, in the spirit of Zeilinger and Brukner, section 6.5). The latter viewpoint is also very supporting for attempts to apply the quantum formalism in cognition and psychology, Chapter 9.

However, for the Copenhagen interpretation the issue of (non)realism has to be taken with caution. Assigning the quantum state to an individual system can be treated as a realistic element of this interpretation. The state dynamics is causal reassembling of the causal state dynamics of Newtonian mechanics. This combination of state realism with the impossibility of the realist representation of randomness of observations is one of the reasons for the popularity of this interpretation. Physicists agree that the quantum experimental situation differs crucially from the situation in classical statistical mechanics: quantum randomness of the results of observations cannot be treated classically. At the same time, even nowadays the majority of experimenters are not ready to give up realism completely. Therefore they are happy to have realism at least at the state level.

Today it is not easy to trace precisely the origin of the orthodox Copenhagen interpretation. I consulted with A. Plotnitsky and I present his reply (which is also important for tracing the origin of another interpretation in the spirit of Copenhagen, the *Copenhagen-Göttingen interpretation*:

"I think that in print the idea appeared first in Dirac's 1927 paper [76] and it might have originated with him. It is not completely clear, however, because Dirac was in Bohr's institute in Copenhagen at the time and he might have gotten it from Bohr. Dirac helped Bohr to translate the *Como lecture* [39], but that was slightly later. Be it as it may on this score, the key to this history is the Como lecture, where Bohr uses this type of interpretation. Here the concept of complementarity was introduced. (The lecture was given in 1927, but first published in 1928, although in his book published in 1931, Bohr gives 1927 as the date of it.) Many, if not most physicists and philosophers, even scholars of Bohr, assumed that Bohr has continued to maintain this view. This is incorrect, because Bohr clearly changed his view beginning with 1928 after his first debate with Einstein. His first publication after this debate in 1929, clearly no longer contains this view. In fact, his position was ambivalent, or perhaps not completely thought through even in the Como lecture. However, he defined complementarity in these terms, as the complementarity between the space-time coordination (observation) and causality (which is unobservable, but assumed). Beginning with 1929 he never speak about complementarity in these terms, and he never says that there is any causality to the independent quantum evolution. In fact he says on many occasions that the application of the idea of causality to quantum processes is out of question. The key complementarities are now those of the position and the application of the momentum conservation law, and time and the application of the energy conservation law, both directly correlative to the uncertainty relations.

However, arguably, still under the impact of the Como lecture this view was adopted by nearly all key early books of the foundations of quantum mechanics, by Heisenberg (in this case, again, ambivalently), Dirac, and von Neumann (also Weyl's book of group theory and quantum mechanics). Pauli appears to give an exception, but he never published a book on an interpretation of quantum mechanics. Many subsequent books (Lahti, Mittelstandt, among other where all seeing it as the Copenhagen interpretation), which adopted this view as the Copenhagen view. This is why the name the orthodox Copenhagen interpretation emerged. I remember talking to Peter Mittelstadt and he told me that Bohr gives a really clear meaning to his interpretation (along this lines) in the Como lecture, especially in German, while Bohr later writing are obscure. I don't think he was right, and more recent scholarship on Bohr, my own included, make it more clear. In fact, this conversation compelled me to consider Bohr's view more carefully. People just don't read or cannot understand Bohr. (Bell, in his criticism of Bohr, was a great example of an unwillingness or inability to read Bohr.) Bohr, for his part, never clearly, if at all, explained this change, perhaps because all his attention was focused on his debate with Einstein, which made him to change his view. Indeed he further refined it after EPR's paper, from which he developed his ultimate interpretation, which I discuss in our article. I explain the changes in Bohr's view in great detail in my book. Heisenberg eventually changes his view as well, but along different lines from Bohr, giving a much greater role to mathematics."

Here Plotnitsky mentioned his book [220]. Thus again by using the personalization approach we can call the orthodox Copenhagen interpretation as the (early)Bohr [39], [40] and Dirac [76], [79], interpretation. However, by taking into account the great contribution of von Neumann in advertising this interpretation through his famous book [250] and his role in clarification of the probabilistic structure of this interpretation, we can call it *Bohr-Dirac-von Neumann interpretation* (but remembering that it matches the views of Bohr only in the period around the Como lecture).

In general, this interpretation says nothing about the corresponding interpretation of probabilities (calculated from multiple observations on quantum systems prepared in the same state ψ). By interpreting probability in different ways one generates various versions of the Copenhagen interpretation.

6.4 Von Neumann's Interpretation

One possibility is to treat probability statistically and it was explored by von Neumann [250]. Therefore it is natural to treat the orthodox Copenhagen interpretation endowed with the statistical interpretation of probability as the *von Neumann interpretation*.

One of the first detailed discussions about the interpretation of probability (and randomness!) in QM was presented in the book [250] of von Neumann who used the orthodox Copenhagen interpretation. In particular, he consistently treated the wave function as representing the state of the concrete quantum system; he was also the first who started the struggle against theories with hidden variables in the formal mathematical framework.

Von Neumann treated quantum randomness as individual randomness, i.e., a single electron is irreducibly random. This interpretation follows naturally from the interpretation of wave function as representing the state of an individual system (e.g., an electron). And, for such an interpretation of the state and randomness, it would be really natural to proceed with the subjective probability, "individual event probability". However, von Neumann proceeded with the statistical interpretation (!) based on the mathematical model of von Mises, frequency probability. The main reason for keeping the statistical interpretation was that aforementioned individual randomness of quantum systems was considered as one of the basic features of nature (and not of human mind!). Von Neumann was sure that such a natural phenomenon must be treated statistically:

"However, the investigation of the physical quantities related to a single object S is not the only thing which can be done – especially if doubts exist relative to the simultaneous measurability of several quantities. In such cases it is also possible to observe great statistical ensembles which consist of many systems $S_1, ..., S_N$ (i.e., N models of S, N large). (Such ensembles, called collectives, are in general necessary for establishing probability theory as the theory of frequencies. They were introduced by R. von Mises, who discovered their meaning for probability theory, and who built up a complete theory on this foundation.)" See [250], p.

298.

We also cite Pauli [215]:

"The theory predicts only the statistics of the results of an experiment, when it is repeated under a given condition. Like the ultimate fact without any cause, the individual outcome of a measurement is, however, in general not comprehended by laws."

The Copenhagen interpretation endowed with von Neumann's treatment of quantum randomness as intrinsic (irreducible randomness) and quantum probability as statistical probability (Mises probability) can be called *von Neumann's interpretation of QM*. Von Neumann emphasized that quantum randomness is a consequence of violation of causality for states of quantum systems. In such a situation it is impossible to provide any kind of realistic description of micro-phenomena. Thus this interpretation is one of a variety of *nonrealist interpretations* of QM, in the sense of our definition of realism, section 6.1.1.

6.5   Zeilinger-Brukner Information Interpretation

In the *information interpretation* of QM, information is the most fundamental, basic entity. Every quantized system is associated with a definite discrete amount of information (see Zeilinger [258], aslo [259]). This information content remains constant during evolution of a closed system. Here a quantum state is defined in the spirit of Schrödinger, see section 6.6: *the quantum state is an expectation catalog (of possible outcomes)*. We remark that Zeilinger elaborated the information interpretation of QM [258] by searching for a fundamental and heuristically clear principle of QM, similar to *Einstein's principle of relativity*.

A. Zeilinger who presented the basic principles of the interpretation in 1999 [258] always emphasized its close connection with the Copenhagen interpretation; in particular, he often cited N. Bohr to emphasize connection with Bohr's ideas. The same line of presentation was continued in joint publications of Zeilinger and Brukner [43] - [45] as well as Kofler and Zeilinger [174]. And really the information interpretation of QM can be considered as a modern information-theoretic version of the orthodox Copenhagen interpretation. It has some commonality with von Neumann's version of this interpretation. In particular, Zeilinger and Brukner explore heavily the concept of *irreducible quantum randomness* which was invented by von Neumann.[4]   However, in contrast to von Neumann, they do not

---

[4] Surprisingly they did not refer to von Neumann at all (of course, it might be that I

consider a quantum state as a state of an individual physical system.

As all interpretations in the spirit of Copenhagen, the information interpretation is non-realistic. We recall that in section 6.1.1 realism was defined in a very general sense: "The ultimate nature of quantum systems and processes is assumed to be representable by means of the mathematical formalism of quantum mechanics or by some subquantum mathematical model." By this interpretation QM is not about features of quantum systems, but about information on these systems gained with the aid of (classical) measurement devices. In section 6.1.1 it was emphasized that one has to distinguish realism and reality. For example, Bohr used the non-realist interpretation of QM, but he definitely does not deny reality of atoms, electrons (and later even photons). It is not clear whether the information interpretation needs reality - beyond outputs of measurement devices (e.g., here photon is treated as a click of a detector).[5]

A. Zeilinger has put forward an idea which *connects the concept of information with the notion of elementary systems*. Here we follow the presentation from his joint paper with J. Kofler [174]:

The description of the physical world is represented by propositions. Any physical object can be described by a set of true propositions. Then Zeilinger pointed out that "we have knowledge or information about an object only through observations."[6] Everybody would agree with this. But further line of reasoning is not unquestionable.

"Any complex object which is represented by numerous propositions can

---

missed some of their papers with the corresponding reference to von Neumann as the inventor of irreducible quantum randomness). They use the term *"objective randomness"*, but the meaning of this term coincides with the meaning of von Neumann's irreducible randomness.

[5]In general Viennese does not like realism and even reality. The following story represents perfectly Viennese's viewpoint on reality. Once my friend Johan Summhammer (Atom Institute, Vienna) spent two weeks in Växjö. This town is surrounded by beautiful lakes ("sjö" is a lake in Swedish). Once I met him staying and looking at a lake, he was really excited. I asked him about the reason of excitement expecting a typical comment about the beauty of Swedish nature. But, Johan answered that he is excited by this great picture created by clicks of detectors composing his brain.' From long conversations with Johan I learned that there is no objective reality, and bio-systems developed an ability to select some patterns from noisy and unstructured environment and cognition was developed in this way (cf. with de Finetti, section 1.11). Thus this world is composed of clicks of detectors. May be not all Viennese share this view of "clicks-made reality", but the above story definitely reflects the general Viennese attitude. There is something in the spirit of the town...

[6]Here one of the important problems of this interpretations lies in assigning the meaning to "we". "Whose knowledge?" as was asked by D. Mermin [201]. We shall come back to this problem later.

be decomposed into constituent systems which need fewer propositions to be specified. The process of subdividing reaches its limit when the individual subsystems only represent a single proposition, and such a system is denoted as an elementary system. The truth value of a single proposition about an elementary system can be represented by one bit of information with "true" being identified with the bit value "1" and "false" with "0". " It is then suggested to assume

Principle of quantization of information: *An elementary system carries one bit of information.*

For me, the main questionable point of this reasoning is validity of such subdivision. Why it is always possible? Can it not be the case that two bits of information belong to a system and cannot be separated? Anyway, the possibility of decomposition as described above looks more like another postulate. From the principle of quantization of information, Zeilinger derives objectivity (irreducibility) of quantum randomness. (In contrast to von Neumann, he needs no mathematically complicated no-go theorem.) Further we again follow the paper of Kofler and Zeilinger [174]:

"Disregarding the mass, charge, position and momentum of the electron, its spin is such an elementary system. If it is prepared "up along z", we have used up our single bit and a measurement along any other direction must necessarily contain an element of randomness. This randomness must be objective and irreducible. It cannot be reduced to some unknown hidden properties as then the system would carry more than a single bit of information. Since there are more possible experimental questions than the system can answer definitely, it has to "guess". Objective randomness is a consequence of the principle lack of information."

I appreciate the elegance of this derivation of irreducibility of quantum randomness. In contrast to von Neumann, Zeilinger need not formulate conditions of coupling between mathematical formalisms of a possible subquantum model and QM. His fundamental principle is formulated in heuristically clear terms, i.e., without any direct relation to, e.g., microworld. However, I stress again that Zeilinger's basic principle of quantization of information is questionable. Other Copenhagenists (Bohr, Heisenberg, Pauli, von Neumann) proceeded without it (see also section 6.6 on "statistical Copenhagen interpretation" of Plotnitsky).

Finally, we remark that in the derivation of irreducibility of quantum randomness there was encrypted the application of the *principle of complementarity* in the form: "Since there are more possible experimental questions than the system can answer definitely...". Thus Zeilinger's information interpretation is based on two fundamental principles:

(1) Principle of quantization of information.
(2) Principle of complementarity.

We note that if two observables (experimental questions) are not complementary, that is, can be asked simultaneously or in any order, it means, mathematically, that their operators commute. There is a theorem, so liked by von Neumann [250], that for commuting operators A and B a new observable C can be constructed such that values of A and B can be unambigously extracted from the result of C measurement - there are two functions $f = f(x)$ and $g = g(x), x \in \mathbf{R}$, such that $A = f(C)$ and $B = g(C)$. Hence, *no complementarity means no inherent randomness*.

Now it is really the time to turn to the questions: "Whose knowledge? Whose information?" I think that these are the "hard questions" of the information approach to QM. Surprisingly here the positions of Zeilinger and Brukner do not coincide.

In June 2014 at the conference in Vienna devoted to 50th anniversary of Bell's inequality, Zeilinger gave the talk presenting the personal agent viewpoint on information encoded in a quantum state. The wave function is in the head and not in nature (von Neumann would definitely disagree, Bohr would probably agree, Wigner would applaud). And the main point is that it is in the *concrete head*, his head or my head. This viewpoint is close to the views of Fuchs, Mermin, Schack, Caves, see section 6.7.[7] This viewpoint on a quantum state as an information entity used in the process of decision making about probabilities of possible outcomes of quantum experiments well matches interpretations of QM as the machinery for update of probabilities - QBism and the Växjö interpretation. I cannot confidently say how Zeilinger interprets probability: statistically or subjectively? Dur-

---

[7]The talk of Zeilinger generated very polarized reactions. For example, Aspect strongly commented that in the two-slit experiment the photon "knows" from the very beginning that it is forbidden to go to some domains on the registration screen, independently of what happens in Zeilinger's head. Then another provocative question was asked to Zeilinger: Is it important to be a human being in order to have a wave function in the head? The reply was in the spirit that this is not important and dog's head is also a good machine to present a wave function! Of course, it might be that this was just a provocative joke response to this provocative question. If not, then additionally to the trouble with Schrödinger cat we got a new trouble - with a Zeilinger's dog. I remark that all these strange creatures were born in Vienna. (But in general this discussion after Zeilinger's lecture reminds the discussion about whether a six-trunk white elephant can have a Buddha nature.)

ing my lectures on foundations of probability at his seminar in May-June 2014 the subjective interpretation of probability was not mentioned at all. I used consistently the statistical interpretation and it seemed that all participants of the seminar were fine with this.

At the same time recently Brukner published a paper on what I interpret as the universal agent perspective on the information interpretation, cf. also with my comments on the views of de Finetti (section 1.11); we cite Brukner [46]:

"The quantum state is a representation of knowledge necessary for a hypothetical observer respecting her experimental capabilities to compute probabilities of outcomes of all possible future experiments."

Here an explicit reference to the observer's experimental capabilities is crucial, cf. QBism, section 6.7.4: *Agents constrained by Born's rule.*

It has to be noted that in this paper the author emphasized the closeness to QBism. With this I strongly disagree. From the QBism perspective, the wave function is in the head of a concrete private agent, e.g., in Fuchs' head, not in the head of a hypothetical observer. Similarly, Zeilinger spoke (at least at the aforementioned occasion) about his private wave function (or even his dog's wave function, again the latter may just be a joke, but later we shall discuss this point seriously), not the wave function used by a hypothetical observer.

The "knowledge" here refers to Wigner's definition of the quantum state: "... the state vector is only a shorthand expression of that part of our information concerning the past of the system which is relevant for predicting (as far as possible) the future behavior thereof."

Unfortunately, it seems that de Finetti and Wigner were not familiar with the works of each other. And the reader can see that de Finetti's viewpoint on probability and Wigner's viewpoint on quantum state match well with each other.

As Brukner pointed out, "Peres correctly notes that considering hypothetical observers is not a prerogative of quantum theory [...]. They are also used in thermodynamics, when we say that a perpetual-motion machine of the second kind cannot be built, or in the theory of special relativity, when we say that no signal can be transferred faster than the speed of light."

This is a natural conclusion; it would be surprising if hypothetical observers and Wigner's interpretation of a state were applicable only in quantum physics. We again refer to de Finetti who did not mention quantum physics at all.

In the light of this statement of Peres and peaceful agreement of the views of de Finetti, Wigner, Zeilinger, Brukner the following question naturally arise: What are the distinguishing features of "information concerning the past of the system which is relevant for predicting" related to quantum systems? Why does information about them represented mathematically in such a special way, by vectors of complex Hilbert space? De Finetti's framework covered homogeneously all areas of science (natural and humanities); in the same way Wigner's viewpoint of a state as a collection of information is applicable to both classical and quantum states of both biological and physical systems. How can the general framework of de Finetti be reduced to the special quantum representation?

One can refer to the principle of quantization of information as the key principle restricting de Finetti's general viewpoint on scientific method; so to say, "classical systems" are those where we are not able to approach the level of a single proposition description, the single bit level. However, this viewpoint suffers from impossibility to derive the quantum (complex Hilbert space) representation from this principle. This problem was understood well by QBists who tried to reconstruct the quantum formalism from their fundamental principle: QM is a special machinery for probability (information) update based on the special nonclassical version of the formula of total probability, section 6.7.3. But QBists succeeded only partially in approaching this great aim. The same problem was actively studied in developing the Växjö interpretation of QM which is based on the same principle as QBism, but explores another version of this formula, section 6.9. The problem of quantum reconstruction was still not solved.

In our opinion to explain peculiarity of the quantum representation of information one has to discuss the logical structure of information processing by humans. Besides classical Boolean logic, there exist a variety of nonclassical logics, nonclassical rules for information processing. In spite of his revolutionary treatment of the concept of probability and the scientific method in general, de Finetti was still rigidly devoted to Boolean logic (and hence to the use of measure-theoretic probability concept). In spite of wide applications of Boolean logic, e.g., in artificial intelligence and computer science, we do not forget that this is just one special model of information processing which was created by a concrete person. Meanwhile, the brain may use more complex logical systems. In particular, it may use *quantum logic*. Thus the quantum representation of information is a mathematical signature of the use of quantum logic in reasoning. Of course, from this viewpoint the context of quantum mechanical reasoning is only a special context in which the brain uses nonclassical logic. Such nonclassical reasoning may be profitable for the brain in other situations, see Chapter 9 on applications of the quantum formalism in cognitive science, psychology, economics; see also [163].

Now we come back to Zeilinger's dog issue. Applying the quantum formalism outside of physics, we discover [163] that nonclassical processing of information is a feature not only of humans, but of all bio-systems, from cells and proteins to, e.g., dogs. From this viewpoint, Zeilinger's dog also processes information (in some situations) by using nonclassical logic of reasoning and hence (roughly speaking) has wave functions in its head.

Of course, the above attempt to couple the information interpretation to general theory of reasoning and decision making is my own speculation; it has nothing to do with the views of adherent of this interpretation (see Chapter 9 for further discussions).

6.6 Copenhagen-Göttingen Interpretation: From Bohr and Pauli to Plotnitsky

This is a non-realist ensemble interpretation, see Bohr [40]. Surprisingly, the best formulation of the Copenhagen-Göttingen interpretation of the quantum state (wave function) can be found in the famous article of Schrödinger [231], where he criticized this interpretation: *the quantum state is an expectation catalog (of possible outcomes).* As in the case of the orthodox Copenhagen interpretation, probability determining expectations can be interpreted either statistically or subjectively. This leads to different versions of the Copenhagen-Göttingen interpretation.

Bohr and Pauli had never formulated their views in the form of a solid and consistent interpretation of QM. Their views about the interpretation of QM also have to be interpreted. As recently proposed by Plotnitsky, e.g., in [223], the *statistical Copenhagen interpretation* (SCI) can be considered as such an attempt. In particular, the concrete interpretation of probability is explicitly incorporated in SCI.

This is a non-realist ensemble statistical interpretation. We present its main points following the work [223]:

a). **Nonrealism** The nature of quantum objects and processes is assumed to be unrepresentable by means of the mathematical formalism of quantum mechanics, or even be inconceivable by human thought. (It is not entirely clear that Bohr goes as far as this last claim, but it is in the spirit of

Copenhagen.) Accordingly, one cannot assign to these processes absolute (irreducible quantum) randomness either. All such features (randomness, correlations, etc., also the uncertainty relations) are only found at the level of measuring instruments.

This allows one to avoid certain contradictory conceptions that appear otherwise, for example, in view of the phenomena observed in the double-slit and other paradigmatic quantum experiments. In this view, such concepts are complementary and, hence, mutually exclusive, but this is not a problem because these phenomena can never be observed in the same experiment or in a set of experiments of the same type (as in the double-slit experiment, in which two different patterns require two alternative and incompatible setups).

On the other hand, one sacrifices the mechanical description of quantum processes or events, including elemental individual ones. This was Einstein's main problem, because it made quantum mechanics incomplete, by his criteria, Einstein-incomplete. The question is, whether it is Bohr-complete, insofar as it is as complete as nature allows our theory to be, at least as things stand now. Einstein believed that a more Einstein-complete theory is possible, while Bohr thought that it might not be possible, which is not the same as saying that it will never be possible.[8]

b). **Noncausality** The absence of causality, even in the case of individual rather than composite processes (as in classical statistical physics), is an automatic consequence of nonrealism. As Schrödinger pointed out, if the classical state does not exist (and it does not, given nonrealism) it cannot change causally, in the sense that the (physical) state of the system at a given point of time determines its states at all later points of time. The absence of determinism, as our inability to merely predict such outcomes, is a weaker claim, because still it may be that quantum processes

---

[8]The latter statement is very important, especially in the light of the present activity on derivation and testing of various no-go theorems. It seems that such activity would not be of much interest to Bohr. (We remark that, in particular, he completely ignored von Neumann's no-go theorem [250]. We suspect similar reaction to the appearance of Bell's theorem.) He was not looking for ultimate "proof" of impossibility to go beyond quantum theory, at least theoretically. For him, both the mathematical structure of QM and experimental research demonstrated that it would be very difficult to create a deeper theory. Presentation of this viewpoint of Bohr on beyond-quantum studies is an important contribution of A. Plotnitsky to quantum foundations. Thus nonrealism of Bohr and SCI is *interpretational nonrealism*. Roughly speaking, this is the most natural and not self-contradictory way to interpret quantum indeterminism, e.g., in the form of Heisenberg's uncertainty relation and interference experiments, and quantum correlations, e.g., in the form of the EPR-correlations.

resulting in such outcomes are causal. Bohmian mechanics is causal (and realist to begin with, or ontological), although it is not deterministic, and its predictions coincide with those of quantum mechanics, and both are in accord with the experimental evidence. The experimental evidence available only implies indeterminism. The lack of causality is an interpretation. This is again an automatic consequence of nonrealism, which, in turn, is due to complementarity and correlations. This interpretation also allows for locality, opposite to Bohmian mechanics.

c). **The Strict Statistical Nature of Quantum Predictions** SCI claims the statistical rather than only probabilistic nature of our predictions, insofar as one cannot, in general, assign probabilities to outcome of individual quantum events. (As such, this is an ensemble interpretation.) Technically, under a) and b) it is still possible to assign, along Bayesian lines, a probability to the outcome of each single individual event, but not according to SCI. In SCI, individual quantum events are, in general, random, but this claim, again, only applies at the level of observed phenomena, as manifested in measuring instruments and not quantum processes themselves.

d). **Information Viewpoint on Quantum State** The quantum state is an expectation catalog (of possible outcomes) [231].

Note that a) and b) are shared (at least to some degree) with Quantum Bayesianism, QBism, but c) may not be, unless the meaning of *objective indeterminism* (see James [120], [122]) is understood in terms of c), while assuming that one could still assign, on whatever grounds (using quantum mechanics or not) a probability to an individual quantum event subjectively. At the same time, some phenomena are statistically correlated, but SCI does not say how this is possible: it offers no concept or history of quantum processes, in contrast to the story of moving objects told by classical physics and relativity. The reason for the statistical approach is, again, that some phenomena, such as those observed in the double-slit experiment, make it difficult to rigorously assign probabilities to individual events, while for collectives one can easily do this. The same is true concerning correlations, which are always statistical.

Now we compare SCI and the von Neumann interpretation. First, von Neumann does not, technically, make the SCI's c) claim. However, this kind of claim may be seen as a consequence or an implication of his statistical view (especially in the above citation about the role of Mises probability in QM), similar to the case of Pauli. Pauli, however, expressly says that individual quantum events are not comprehended by physical laws (which, since he does not say otherwise, may be assumed to include the probabilistic laws of quantum mechanics, established through the wave function and Born's or related rules, such as von Neumann's projection postulate). This type of claim, i.e., impossibility of assigning probabilities to individual quantum events, cannot be found in von Neumann's writings.

Secondly, von Neumann's basic principle is violation of causality. And nonrealism is a straightforward consequence of this violation. Plotnitsky proceeds the other way around. For him, the basic principle is nonrealism.

SCI is one of the strongest versions of the nonrealistic interpretation of QM. As was pointed out above, it is even not clear whether Bohr by himself would sign under this statement. This declaration of nonrealism is so important that we analyze it in more detail. It contains two parts which carry the messages with sufficiently different, although closely related, contents. The first message is that quantum observables and states cannot be treated in the realist way (for Einstein, Podolsky and Rosen [85] this proved incompleteness of QM): for example, the position and momentum values cannot be assigned to an electron in the same way as they are assigned to molecules of a classical gas. The second message is that *it is impossible to construct any subquantum model which would reproduce statistical predictions of QM.*

6.7   Quantum Bayesianism - QBism

6.7.1   *QBism childhood in Växjö*

In 2001 QBism was strongly represented at the second Växjö conference on quantum foundations, "Quantum Theory: Reconsideration of Foundations" (QTFT2001), June 17-21, 2001. We (organizers and participants of this conference) strongly believed that the quantum information revolution would soon lead to great foundational revolution. Unfortunately, dreams did not come true. Nevertheless, the energy of the quantum information revolution was transformed in a series of stormy debates during the series of the Växjö conferences, 2000-2015. Although these debates did not lead to a complete resolution of the basic problems of quantum mechanics, they clarified some of these problems, especially the problem of the interpretation of a quantum state. QBism was definitely one of the main foundational outputs of the quantum information revolution.[9]

---

[9]Besides QBism, we can mention the *Växjö interpretation* of QM (statistical realist and contextual) [143], [147] derivation of the QM-formalism from simple operational

As the organizers of QTFT-2001, I and C. Fuchs both dreamed for creation of a consistent and clear interpretation of QM, free of mysteries and paradoxes. However, we went in two opposite directions. I followed Einstein and later, as a result of better understanding of Bohr's writings (and especially comments of A. Plotnitsky on them), tried to unify Einstein's realist statistical interpretation with Bohr's contextual interpretation by filtering out Bohr's nonrealist attitude, see section 6.9 for so called *Växjö interpretation* of QM. Both Einstein and Bohr (as well as, e.g., von Neumann) used the statistical (ensemble) interpretation of quantum probabilities. Therefore the Växjö interpretation is based on this statistical interpretation. It was also important that as student I was strongly influenced by A. N. Kolmogorov and B. V. Gnedenko who always emphasized that probability is objective and statistical, see also section 1.12. Later, after PhD, I discovered works of von Mises. I really enjoyed this reading! Von Mises' interpretation of probability differs from Kolmogorov's interpretation, see Chapter 1. Nevertheless, Misesian probability is also objective and statistical. In any event, nobody of them (Einstein, Bohr, von Neumann, Kolmogorov, Gnedenko) and neither I would agree with De Finetti's slogan: *"Probability does not exist!"* and with his subjective interpretation of probability. My views on interpretation of quantum states and probabilities were presented in [143].

C. Fuchs went in the opposite direction, he (with the support of C. M. Caves, R. Schack, and D. Mermin) openly, loudly, and proudly declared [52], [53], [93] - [99] that QM is only about knowledge (and here QBists are very close to the fathers of the Copenhagen interpretation, N. Bohr, and W. Heisenberg). But this widely supported viewpoint was completed with very strained and revolutionary declaration that this quantum knowledge has to be treated as *personal knowledge*. Subjective interpretation of quantum probability matches perfectly such a private agent perspective of quantum theory. QBists emphasize the Bayesian probability update and decision making structure of the quantum probability calculus.

From my viewpoint, the latter is one of the main contributions of QBism in clarification of quantum foundations. Independently this viewpoint was presented in the framework of the Växjö interpretation, section 6.9.

At the beginning I and C. Fuchs did not recognize this similarity, namely,

---

principles, D' Ariano [63], [64] and Chiribella et al. [57], [58] (first time this project was also announced in Växjö), and the *statistical Copenhagen interpretation* (statistical non-realist) which final formulation was presented at the Växjö-15 conference by A. Plotnitsky.

interpretation of the calculus of quantum probabilities as a machinery for update of probabilities. And it is clear why: at that time I emphasized realism and objectivity and C. Fuchs privacy and subjectivity. And the probability update dimension was shadowed by these philosophic issues. This explains the appearance of Fuch's anti-Växjö paper [94].

Nevertheless, intuitively I felt sympathy to QBism but roots of this sympathy were not clear for me.[10] At the Växjö-15 conference QBism was widely represented and celebrated its worldwide recognition. Nobel Prize Laureate T. Hänsch presented the great lecture about QBism as the only possible consistent foundational basis of quantum information theory. This lecture ignited the stormy debate and T. Hänsch and C. Fuchs were attacked by the realist opposition (led by L. Vaidman and A. Elitzur).

6.7.2 Quantum theory is about evaluation of expectations for the content of personal experience

In contrast to von Neumann, Fuchs proposed to interpret probability in the subjective way. To present essentials of some theory, sometimes it is practical simply to cite works of its creators (this is definitely not the case of Bohr, or von Neumann, or even myself). Here we cite C. Fuchs and R. Schack [98], pp. 3-4:

"The fundamental primitive of QBism is the concept of experience. According to QBism, quantum mechanics is a theory that any agent can use to evaluate her expectations for the content of her personal experience.

QBism adopts the personalist Bayesian probability theory pioneered by Ramsey [...] and de Finetti [...] and put in modern form by Savage [...] and Bernardo and Smith [...] among others. This means that QBism interprets all probabilities, in particular those that occur in quantum mechanics, as an agent's personal, subjective degrees of belief. This includes the case of certainty - even probabilities 0 or 1 are degrees of belief. ...

In QBism, a measurement is an action an agent takes to elicit an experience. The measurement outcome is the experience so elicited. The measurement outcome is thus personal to the agent who takes the measurement action. In this sense, quantum mechanics, like probability theory, is a single user theory. A measurement does not reveal a pre-existing value. Rather, the measurement outcome is created in the measurement action.

According to QBism, quantum mechanics can be applied to any physical system. QBism treats all physical systems in the same way, including atoms, beam

---

[10]Once Christopher Fuchs asked me: "Why did you support QBism so strongly during the Växjö-series of conferences? QBism contradicts your own Växjö interpretation!" In fact, I was not able to explain this even for myself. I had a feeling that QBIsm can be useful. But how? and where? see Chapter 9.

splitters, Stern-Gerlach magnets, preparation devices, measurement apparatuses, all the way to living beings and other agents. In this, QBism differs crucially from various versions of the Copenhagen interpretation. ...

An agent's beliefs and experiences are necessarily local to that agent. This implies that the question of nonlocality simply does not arise in QBism."

Remark 1. Here is a good place to comment on the QBist statement that "the case of certainty - even probabilities 0 or 1 are degrees of belief." I agree that in applications of probability these probabilities are typically treated as representing the case of certainty, "never happen" and "always happen". However, this viewpoint does not match the mathematical models of probability. In the Kolmogorov measure-theoretic approach it is evident that an event A with $p(A) = 0$ can happen. It is just "practically never happen". And here "practically" refers to the concrete context and concrete decision maker. Moreover, let us turn to the Kolmogorov interpretation of probability, section 1.8, its b-part: "if $P(A)$ is very small, one can be practically certain that when conditions Σ are realized only once the event A would not occur at all." Thus, for Kolmogorov, zero probability does not differ so much (from the practical viewpoint) from "very small probability". The same can be said about von Mises theory. Thus I think that the proper treatment of probabilities 0 or 1 in the statistical and subjective theories of probability does not differ so much.

We shall revisit the interpretational issues of QBism after the presentation of its basic probabilistic principle in the next session.

6.7.3   QBism as a probability update machinery

The previous section might create the impression that the subjective interpretation of quantum probabilities is the key point of QBism. It might be that even its creators have the same picture of their theory. For me, the essence of QBism is neither this very special interpretation of quantum probabilities nor the concrete agent perspective of QM. For me, the main ideological invention of C. Fuchs and R. Schack was treatment of the mathematical formalism of QM as a generalization of the classical Bayesian machinery of the probability update. This viewpoint clarifies the meaning of the basic rule of QM – the Born rule as a complex Hilbert space representation of generalization of the *classical formula of total probability* (FTP).

This dimension of QBism is identical to the probability update dimension of the Växjö interpretation, section 6.9. There are a few differences.

One is the difference in the interpretations of probability. However, nowadays I do not consider it as the crucial difference (in contrast to my first debates with C. Fuchs in 2001-2003, [143] and [94]). Really, in a long series of updates the subjective and statistical viewpoints coincide.[11]

For me, the main difference between QBism and the Växjö interpretation is in the mathematics, not in physics or in philosophy. As we shall see in section 6.9, by both the Växjö interpretation of QM [141], [142], [161] and QBism [96] - [99] *the Born rule is treated as generalization of classical formula of total probability* (FTP) in the language of linear operators. However, these interpretations are based on two totally different mathematical generalizations of FTP - both matching the Born rule.

Starting with the Born rule, QBists derived their special version of generalized FTP which is based on a very special class of the quantum probability updates, based on atomic instruments with SIC-POVMs (section 4.6).

We now briefly present the QBism scheme for the probability update, namely, the representation of the Born rule as a generalization of FTP; here we again follow Fuchs and Schack [96] - [98].[12]

Quantum states are represented by density operators ρ in a Hilbert space assumed to be finite dimensional. A measurement (an action taken by the agent) is described by a POVM $F = (F_j)$, where j labels the potential outcomes experienced by the agent. The agent's personalist probability[13]

---

[11] In the mentioned debates I was also strongly against the anti-realist attitude of QBism. However, now this attitude does not disturb me as much as 12 year ago. Either I started to understand QBists views on the problem of realism better or QBists changed their views (or both). QBism needs not appeal to any subquantum model providing the ontic description of quantum systems and processes, in particular, to hidden variables. Nor is QBism concerned with struggle against such models. It seems that the personal position of C. Fuchs is similar to the position of N. Bohr: for quantum *physics*, it plays no role whether finally one would be able to construct a realistic subquantum model or not. For the present state of development of quantum theory, this is the most reasonable position, cf. with Zeilinger's strong anti-realist attitude. Moreover, just recently (through a series of email exchanges) I understood better the position of C. Fuchs on the problem of non/realism. Surprisingly QBists (at least C. Fuchs) do not consider QBism as a non-realist interpretation of QM. As was already remarked, their viewpoint on no/realism differs crucially from our mathematical modeling viewpoint.

[12] We remark that in coming considerations the interpretation of probabilities does not play any role. They can be subjective probabilities (as originally in QBism), but they can also be statistical, e.g., Kolmogorovian or Misesian, as well.

[13] As was remarked, probability can be interpreted in other ways. The situation is similar to the classical probability update. De Finetti would treat this probability as subjective, but Kolmogorov, or Gnedenko, or von Mises as statistical (Chapter 1).

$p(F_j)$ of experiencing outcome j is given by the Born rule:

$$p(F_j) = \mathrm{Tr} F_j \rho. \tag{6.1}$$

Similar to the probabilities on the left-hand side of the Born rule, QBism regards the operators ρ and F_j on the right-hand side as judgments made by the agent, representing her personalist degrees of belief.

The Born rule as written in Eq. (6.1) appears to connect probabilities on the left-hand side of the equation with other kinds of mathematical objects - operators - on the right-hand side.

QBists assume that the agent's reference measurement is an arbitrary *informationally complete POVM*, $E = (E_i)$, such that each E_i is of rank 1, i.e., is proportional to a one-dimensional projector. Such measurements exist for any finite Hilbert-space dimension (section 4.6). Furthermore, we assume that, if the agent carries out the measurement $E = (E_i)$ for an initial state ρ, upon getting outcome E_i he would update to the post-measurement state

$$\rho_i = \frac{E_i \rho E_i}{\mathrm{Tr} E_i \rho E_i}.$$

This is the assumption of atomicity of this quantum instrument (section 4.5). (This is a strong constraint on the class of instruments which are used in QBism. It would be interesting to analyze the possibility to proceed with arbitrary instruments.)

Because the reference measurement is informationally complete, any state ρ corresponds to a unique vector of probabilities

$$p(E_i) = \mathrm{Tr} E_i \rho,$$

and any POVM $F = (F_j)$ corresponds to a unique matrix of conditional probabilities

$$p(F_j|E_i) = \mathrm{Tr}(F_j \rho_i).$$

The operators ρ and F_j on the right-hand side of the Born rule are thus mathematically equivalent to sets of probabilities $p(E_i)$ and conditional probabilities $p(F_j|E_i)$. (We remark that all these probabilities depend on the state ρ, i.e., $p(E_i) \equiv p_\rho(E_i), p(F_j) \equiv p_\rho(F_j), p(F_j|E_i) = p_\rho(F_j|E_i)$.)

Then Fuchs and Schack [96] - [98] stress that POVMs as well as quantum states represent an agent's personal degrees of belief. However, this is not essential for the formal scheme of probability update. We can as well interpret the probabilities $p(E_i)$ and $p(F_j|E_i)$ statistically. The main point which was rightly emphasized by them is that the Born rule can be interpreted as one special form of transformation of probabilities:

$$p(F_j) = f(p(E_i), p(F_j|E_i)). \tag{6.2}$$

Comparing with the classical FTP which is in these notations written as

$$p(F_j) = \sum_i p(E_i)p(F_j|E_i) \tag{6.3}$$

they formulate the statement which I consider as the cornerstone of QBism:

The Born rule is one of the forms of generalization of FTP.

For me, the main problem of QBists is that they started with a SIC-POVM $E = (E_i)$. They say [96]: "The Born rule allows the agent to calculate her outcome probabilities $p(F_j)$ in terms of her probabilities $p(E_i)$ and $p(F_j|E_i)$ defined with respect to a counterfactual reference measurement." This reference to counterfactuals is really redundant. Why should SIC-POVM measurement appear at all? As in the Växjö approach (section 6.9), one can start with an arbitrary POVM-measurement, say $G = (G_i)$, to define probabilities $p(G_i) = \mathrm{Tr}\rho G_i$ providing information about the state ρ.

By taking into account such a possibility of generalization of QBist consideration[14] I completely agree with the following statement of QBism [96]:

"In QBism, the Born rule functions as a coherence requirement. Rather than setting the probabilities $p(F_j)$, the Born rules relates them to those defining the state ρ and the POVM $F = (F_j)$. Just like the standard rules of probability theory, the Born rule is normative: the agent ought to assign probabilities that satisfy the constraints imposed by the Born rule."

The functional relationship given by (6.2) depends on details of the reference measurement. In the special case that the reference measurement is a symmetric informationally complete POVM (SIC), (6.2) takes the simple form:

$$p(F_j) = \mathrm{Tr}F_j\rho_i = \sum_i \left((d+1)p(E_i) - \frac{1}{d}\right)p(F_j|E_i). \tag{6.4}$$

This is a consequence of the complete information representation of a quantum state and the use of SIC-POVMs (section 4.6, Eq. (4.44)).

---

[14]However, for QBists the above generalization - to start the probability update scheme with an arbitrary POVM measurement $G = (G_j)$ and not with a SIC-POVM $E = (E_i)$ - seems to be unacceptable. They are really addicted on SIC-POVMs and on completeness of information gained at the first step, information about the state, even at the price of appearance of counterfactuals.

QBists have conjectured that this form of the Born rule may be used as an axiom in a derivation of quantum theory. The key question that remains is in identifying what minimal further principles must be added to Eq. (6.4) for the project to be successful.

Remark 2. This program is identical to attempts to justify the Växjö interpretation (section 6.9) by deriving the complex Hilbert space formalism from the generalized FTP with an interference term (6.8). However, I tried to derive the complex Hilbert space structure of QM solely from the latter generalization of FTP, i.e., without additional axioms (Chapter 5). This approach was successful only for dichotomous observables. Already the case of three-valued observables is very difficult mathematically. Here only a partial success was achieved by P. Nyman and I. Basieva [207], [208]. May be Fuchs and Schack are right that one has to find additional axioms leading to the complete derivation of the quantum formalism, as, e.g., was done by D' Ariano et al. [63], [64], [57], [58]. However, proceeding with such additional axioms does not match completely the basic principle that the quantum formalism is just a special form of the probability update generalizing the classical Bayesian update. This principle is very attractive in both QBism and the Växjö interpretation (though they have different mathematical realizations, (6.4) or (6.8)). Completing a generalized probability update scheme by additional operational principles diminishes the value of this scheme as the *unique fundamental principle* lying in the ground of QM. It shifts the line of research to more traditional operational approaches starting with the pioneer contribution of Heisenberg [108], then Mackey [195] and nowadays D' Ariano et al. [63], [64], [57], [58]. In this book we do not plan to analyze the operational axiomatics of D'Ariano et al. However, in previous derivations of QM from "natural operational principles" complex Hilbert space was always encrypted, practically explicitly, in one of the axioms, see, e.g., G. Mackey [195] for one of the first operational derivations of the quantum formalism. There is still some hope that at least the Växjö scheme (section 6.9) of the probability update, for the FTP with the interference term (6.8) may lead to derivation of the complete formalism. The difficulties which stopped my further studies in this direction were of the mathematical origin – too complicated mathematical calculations.

6.7.4 Agents constrained by Born's rule

In 2001 the private user's experience viewpoint on quantum *physics* made me really mad and this was the main reason for my anti-QBism attitude [143] (see Appendix A).[15] However, recently I understood that the situation may be not so bad as one can imagine by reading the QBism manifests, such as presented in section 6.7.2. May be this long way to understanding is not only my fault. QBists judge too highly the private user interpretation of QM comparing with the problem of concretization of the class of such private users. I remember that in 2001 in Växjö I asked C. Fuchs: "Suppose that your user, Ivan, lives in taiga by hunting and he has never heard about QM. Would Ivan make proper predictions about simplest quantum experiments?" I do not remember the precise answer of C. Fuchs, but it seems it was a long story about his version of FTP (6.4), which was considered by me as totally irrelevant to my question. Then I asked C. Fuchs the same question during a few next Växjö conferences and I did not get a satisfactory answer (at least from my viewpoint). Recently, during the Växjö-2015 conference, where QBism was heavily represented by the talks of its founders, I got a new possibility to discuss its foundations with C. Fuchs and in the after-conference email correspondence I finally got a clear answer to my old question. Of course, this answer could be found in the works of Fuchs and Schack, but it was dimmed by the very strong emphasize of the private agent perspective.

So, if I understood C. Fuchs correctly, the class of agents has to be *constrained!* And the basic constraint is given by the Born rule which is treated as an empirical rule reflecting some basic features of nature, see section 6.7.3 for discussion. For a moment, for us the concrete natural basis of the Born rule is not important. It is important that QBism uses this rule as an information constraint to determine a class of so to say "quantum agents", i.e., those who "get tickets to the QBism performance." Thus private users of QM are those who know the main rule of the game: *the probability update for quantum systems has to be done with the aid of the Born rule (or QBist version of FTP, see (6.4)).* It seems reasonable that such agents would produce reasonable predictions. Thus Ivan from taiga is excluded from QBist agents – finally!

---

[15]The strong anti-Copenhagen attitude in the first declaration about the Växjö interpretation was partially a consequence of the active advertising of QBism at Växjö-2001 conference. My reaction (as many others) was: "See, the Copenhagen interpretation finally led to such a perverse view on QM as the private agent's perspective on the quantum state."

The Born rule constraint is the basic necessary condition for entrance to the QBism club. At the same time it is practically a sufficient conditions, because other personal characteristics of an agent making predictions about quantum experiments play subsidiary roles in relation to these predictions. Thus in principle one may invent an abstract (conceptual) QBism-agent who makes her probabilistic predictions about experiment results on the basis of the Born rule. In this way QBism comes closely to the recent version of the information interpretation of QM of Zeilinger-Brukner proposed by C. Brukner [46], see section 6.5. However, while QBists would, in principle, accept an interpretation a la Brukner, i.e., referred to a conceptual agent, they definitely would not like to diminish the role of private agent perspective in QBism.

Finally, I point to coupling of QBism with recently flowering applications of the probabilistic formalism of quantum mechanics outside of physics, see Chapter 9. C. Fuchs and R. Schack declared [98]: "According to QBism, quantum mechanics can be applied to any physical system. QBism treats all physical systems in the same way, including atoms, beam splitters, Stern-Gerlach magnets,..." If so, then why only "to any physical system"? Why not to any system, biological, cognitive, social, political? From my viewpoint, QBism is an excellent interpretation to motivate extension of the domain of applications of the quantum formalism.

However, it seems that, for a QBist, it is difficult (if possible at all) to accept the possibility of such wider use. The reaction of C. Fuchs to my comments in this direction cannot be characterized as excitement. Of course, this calm reaction might have social roots. By assuming that QBism is a theory about generalized probability update done not only in physics, but, in fact, everywhere, QBists would depart even farther from the main stream physics.

However, it might be that distancing from applications outside of physics has fundamental grounds. In contrast to the Växjö interpretation (section 6.9), in QBism the Born rule is not just a consequence of a very general scheme of the probability update. Its appearance in quantum physics is a consequence of some fundamental feature of nature, namely, a kind of intrinsic quantum randomness. Thus by extending the domain of applications of the quantum formalism to cover, e.g., cognition, one has to assign to cognition a kind of intrinsic quantum(-like) randomness. In principle, one cannot exclude that not only quantum physical systems, but even bio-systems are intrinsically random. However, this is a very complicated problem. It seems that QBists (busy with their own problems in physics) simply do not like to be involved in the problem of justification of intrinsic bio-randomness.

6.7.5 QBism challenge: Born rule or Hilbert space formalism?

This section is based on the output of my discussions with A. Plotnitsky on the QBists emphasis of the role of the Born rule in QM and their attempts to derive it from the generalized scheme for probability update based on the QBist-FTP (6.4).

In QBism the Born rule has very high esteem. It is important to point out that this rule is just a component of the complex Hilbert space representation of states of quantum systems (independently of their interpretation), i.e., it is just a part of the *mathematical model* serving QM. If the complex Hilbert space representation is established (derived or postulated), then Born's rule appears practically automatically:

The wave function cannot be used directly to predict probabilities, because it is a complex vector in a Hilbert space, and you need a real number to get to probabilities, for which the conjugation is the easiest way, which is the Born rule.

Heisenberg already de facto used it for the special case of the probabilities for elections emitting photon in the hydrogen atom. Dirac and von Neumann, of course, used in connection with the projection postulate. Hence, given the Hilbert space representation of states, the Born rule is its trivial consequence.

In QBism the Born rule is interpreted as an empirical rule. However, the Born rule is not simply a feature of nature, but rather, along with the mathematical model itself[16], the rule responds to something that is in nature and in our interactions with nature, which is, in a way, also part of nature, insofar as we are part of nature. It is also important to remark that the Born rule is added to the Hilbert space formalism and not derived from it, but both are always used together for any prediction.

We also remark that, in contrast to QBism, the Växjö approach (section 6.9) can be characterized as an attempt to derive the Hilbert space representation of probabilities with aid of complex probability amplitudes (wave

---

[16]They always work together forming a coherent set of constraints upon effective predictions, as the Born rule is just as meaningless apart from the mathematical model as the use of the mathematical formalism cannot predict the outcomes of measurements without Born's rule, see footnote 22 of paper [223].

functions), not just the Born rule in, e.g., SIC-POVM form. This assigns a higher value to the wave function Hilbert space representation than in QBism. This is a special case of the mathematical modeling approach to study of natural (and mental) phenomena, again cf. section 6.9.

This question can be considered from even more general point: even if we have a different mathematical model, somehow without amplitudes (wave functions), but still with some Born or Born-like rule (for example, because the SIC-POVM formalism is still complex-number based), the agent's commitment to this rule will necessarily cover the mathematical model, too. Arguably, it will be to the model in the first place, because the rule applies in the framework of the model and is meaningless without it. For example, one needs to be committed to the mathematics which defines SIC-POVMs, if only in a trivial sense of manipulating the formal symbols involved. On the other hand, we need such a rule to make predictions in cases like QM.

In classical mechanics, or relativity, the formalism is descriptive and predictions follow from this description, so the rule is contained in the formalism.

May be QBism current program is to embed the rule in the mathematical model somehow, but it is not clear whether this is possible, as the QBism-formalism is clearly not descriptive.

That is not only the question of adding probability calculation (the rule for how amplitudes change), but how to get to probability in the first place. It was not easy even in classical statistical physics, and it took the likes of Boltzmann, Maxwell, Planck, and Gibbs, to establish these rules.

6.7.6   *QBism and Copenhagen interpretation?*

By following the talks of C. Fuchs or R. Schack I always had the feeling that their views are very much in the spirit of Copenhagen, especially its Göttingen-Copenhagen version, section 6.6. Bohr and Heisenberg always pointed out that the quantum formalism is not about the "quantum physical world", but it is a representation (mathematical) of measurements performed on micro-systems. Thus, from their viewpoint QM is about knowledge (especially for Heisenberg). Do Fuchs and his coauthors, Schack, Mermin, Caves, try to say the same thing by just using the special interpretation of probability, the subjective probability? I still do not have my own definite opinion about this issue. Therefore here I present a long citation of D. Mermin who knows QBism much better than me and who claims that

QBism differs crucially from all interpretations in the spirit of Copenhagen, see [202], pp. 7-8:

"A fundamental difference between QBism and any flavor of Copenhagen, is that QBism explicitly introduces each user of quantum mechanics into the story, together with the world external to that user. Since every user is different, dividing the world differently into external and internal, every application of quantum mechanics to the world must ultimately refer, if only implicitly, to a particular user. But every version of Copenhagen takes a view of the world that makes no reference to the particular user who is trying to make sense of that world.

Fuchs and Schack prefer the term "agent" to "user". "Agent" serves to emphasize that the user takes actions on her world and experiences the consequences of her actions. I prefer the term user to emphasize Fuchs' and Schack's equally important point that science is a user's manual. Its purpose is to help each of us make sense of our private experience induced in us by the world outside of us.

It is crucial to note from the beginning that "user" does not mean a generic body of users. It means a particular individual person, who is making use of science to bring coherence to her own private perceptions. I can be a "user". You can be a "user". But we are not jointly a user, because my internal personal experience is inaccessible to you except insofar as I attempt to represent it to you verbally, and vice-versa. Science is about the interface between the experience of any particular person and the subset of the world that is external to that particular user. This is unlike anything in any version of Copenhagen. It is central to the QBist understanding of science."

Of course, the reader can find this Mermin's viewpoint on QBism as private agent (user) business does not match completely the conclusion from my discussion on the class of users belonging to the "QBism club" and constrained by the knowledge about the Born rule, see section 6.7.4. This is a consequence of the private user's perspective to interpreting QBism – both David Mermin and I are good friends of the founder of QBism, Christopher Fuchs, and we both got our information about QBism directly from its founder....

6.8 Interpretations in the Spirit of Einstein

A. Einstein who contributed a lot in mathematical justification of classical stochastic phenomena, including theory of Brownian motion, of course, wanted to keep statistical interpretation of probability for QM, too; more concretely, he wanted to continue to use the measure-theoretic framework.[17]

---

[17]As was noted in Chapter 1, Kolmogorov's formalization of classical probability came too late to play any role at the initial stage of creation of QM. However, the majority of

By Einstein's interpretation a quantum state (wave function) represents statistical properties of an ensemble of identically prepared quantum systems. Up to this point, Einstein's interpretation does not differ from von Neumann's and Pauli's interpretations (as well as from SCI of Plotnitsky). However, in contrast to people inspired by Copenhagen, Einstein considered these observable statistical properties as *objective properties of systems* (ontic properties).

In short, Einstein's interpretation is the realist ensemble statistical interpretation. It does not differ from the interpretation of classical statistical physics. Ballentine put tremendous efforts to advertising and clarification of this interpretation [25], [26]. He used the term *"statistical interpretation"* of QM. However, this terminology can be misleading, because adherent of the Copenhagen interpretation using the statistical interpretation of probability, e.g., von Neumann and Pauli (and it seems also Bohr) also treated their interpretations as statistical (as well as Plotnitsky does nowadays).

The main problem of Einstein's interpretation was sharply formulated in a letter from Schrödinger to Einstein [214]. Schrödinger was very sympathetic to the realist ensemble statistical interpretation of QM and he expressed his sympathy in this letter. However, then he pointed out (with sadness) that he could not keep this interpretation for quantum phenomena (although it is perfect for classical statistical mechanics), because it cannot explain the interference phenomenon. In other words this interpretation definitely confronts complementarity and quantum uncertainty. If we analyze the real roots of Schrödinger's dissatisfaction, it becomes clear that he wanted a proper interpretation of QM explain the origin of the Born rule, the (physical?) meaning of wave functions. However, the interpretations in the spirit of Copenhagen failed at this much the same as the statistical interpretation did. Therefore in the same letter Schrödinger also expressed his dissatisfaction by the Copenhagen interpretation. And the famous and nowadays overexploited example with *Schrödinger cat* was elaborated just to show absurdness of this interpretation. We remark that Schrödinger's "cat and poison" metaphor is just a modification of the original Einstein's "man and gun" metaphor which was presented in one of Einstein's letters to Schrödinger. However, Einstein's "dead-alive man" did not become fa-

---

scientists had already used theory of measure to define probability. In contrast to later Kolmogorov's abstract and very general measure-theoretic model, they (e.g, Borel and Frèchet) used *concrete measures and sample spaces* for concrete probability problems, see [156]. Kolmogorov was able to separate the probabilistic and geometric contents. For him it was not important whether events were represented by subsets of real line (Borel) or complete metric space (Frèchet).

mous, in contrast to Schrödinger's "dead-alive cat." And it is clear why. By taking into account clearly expressed Einstein's attitude against the Copenhagen interpretation, it would be impossible to explore the "man and gun" metaphor without pointing to absurdness of the Copenhagen interpretation. Schrödinger, although not sympathetic to Copenhagen interpretation, was not actively struggling against it. Therefore it was possible to use the "cat and poison" metaphor to emphasize mystical features of quantum systems - without mentioning the real origin of this metaphor.

6.9 Växjö Interpretation

The Växjö interpretation [143], [146], [147] is the *realist statistical contextual interpretation* of QM. Thus, in contrast to Copenhagenists (from Bohr, Heisenberg, and Pauli to Plotnitsky)[18] or QBists, by the Växjö interpretation QM is not complete and it can emerge from a subquantum model. This interpretation is statistical, i.e., probability is endowed with the statistical interpretation - due to von Mises or Kolmogorov (Chapter 1). In this book we proceed in Kolmogorov's framework, see [156] for detailed presentation of the Växjö approach in von Mises' framework. My interpretation is contextual, i.e., experimental contexts have to be taken into account really seriously (Chapter 5).[19]

By the Växjö interpretation the probabilistic part of QM is a special mathematical formalism to work with contextual probabilities for families of contexts, which are, in general, incompatible. Of course, the quantum probabilistic formalism is not the only possible formalism to operate with contextual probabilities (see Chapter 5).

The main distinguishing feature of the formalism of quantum probability is its complex Hilbert space representation and the Born rule. *All quantum contexts can be unified with the aid of a quantum state ψ* (wave function,

---

[18]Sometimes QBists and adherents of the information interpretations of QM, e.g., Zeilinger, Brukner, Bub are classified as "neo-Copenhagenists". However, as was stressed by Mermin (section 6.7.6), QBism cannot be treated as a modification of the Copenhagen interpretation. And from my viewpoint information-like interpretations are so far from the original Copenhagen interpretation, e.g., von Neumann's interpretation, that I am completely sure that adherents of such interpretations would not be welcome to the traditional Copenhagen club.

[19]In my numerous talks I experienced (from the reaction of the audience) that people typically are not comfortable with combination of realism and contextuality. However, as we can see, for example, in PCSFT (see section 5.2.2), the classical random field model of quantum phenomena, contexts can be represented by using the causal space-time picture.

complex probability amplitude). A state represents only a part of context, the second part is given by an observable A. Thus the quantum probability model is not just a collection of Kolmogorov probability spaces $\mathcal{P}_C$.

In a general contextual probabilistic model, each context $C \in \mathcal{C}$ is represented by its own Kolmogorov probability space $\mathcal{P}_C$ endowed with its own probability measure P_C and, moreover, with its own space of random variables R_C representing observables. In spite of the above statement that the quantum probabilistic formalism is only a very special contextual probabilistic formalism, there are some arguments in favor the hypothesis that "natural contextual probabilistic models" can be represented in the quantum-like way, i.e., in complex Hilbert space or its generalizations based on other algebraic systems [161] (Chapter 5). The same hypothesis plays the fundamental role in justification of QBism and, as I know, C. Fuchs, R. Schack, I. Begtsson put tremendous efforts to solve this problem - for a moment, their success is only partial.

Each theory of probability has two main purposes: descriptive and predictive. In classical probability theory its predictive machinery is based on *Bayesian inference* and, in particular, the *formula of total probability* (FTP) (see section 1.6).

Can the probabilistic formalism of QM be treated as a generalization of Bayesian inference?

My viewpoint is that quantum FTP (see section 5.6) is, in fact, a modified rule for the probability update. QM provides the following inference machinery. There are given a state ρ and two observables $A = (M_i^a = V_i^\star V_i)$ and $B = (M_i^b = U_i^\star U_i)$. The observables are given by POVMs corresponding to quantum apparatuses - the instruments of the atomic type (the latter restriction is made just to simplify considerations). The first measurement of A can be treated as collection of information about the state ρ. The result $A = \alpha_i$ appears with the probability

$$p^a(\alpha_i) = \mathrm{Tr} M_i^a \rho \qquad (6.5)$$

and the initial state ρ is transferred to the state

$$\rho_{a_i} = \frac{V_i^\star \rho V_i}{\mathrm{Tr} V_i^\star \rho V_i}. \qquad (6.6)$$

Then, for each state ρ_{a_i}, we perform measurement of B and obtain probabilities

$$p(\beta_j | \alpha_i) = \mathrm{Tr} M_j^b \rho_{a_i}. \qquad (6.7)$$

We now recall the quantum FTP (5.91) (section 5.6.2):

$$p(b = \beta_j) = \sum_k p(b = \beta_j | a = \alpha_k, \bar{a} = \alpha_k) p(\bar{a} = \alpha_k | a = \alpha_k) p(a = \alpha_k)$$

$$+2 \sum_{k<m} \cos \phi_{j;km}$$

$$\times \sqrt{\prod_{i=k,m} p(b = \beta_j | a = \alpha_i, \bar{a} = \alpha_i) p(\bar{a} = \alpha_i | a = \alpha_i) p(a = \alpha_i)}. \quad (6.8)$$

Thus we can predict the probability of the result β_j for the b-observable on the basis of the probabilities for the results α_i for the a-observable and conditional probabilities. In the case of observables of the von Neumann-Lüders type, the probability update rule is considerably simplified (equality (5.85), section 5.6.1):

$$p(b = \beta) = \sum_k p(b = \beta | a = \alpha_k) p(a = \alpha_k) \quad (6.9)$$

$$+2 \sum_{k<m} \cos \phi_{j;k,m} \sqrt{p(b = \beta | a = \alpha_k) p(a = \alpha_k) p(b = \beta | \alpha_m) p(a = \alpha_m)}.$$

Of course, the main nonclassical feature of this probability update rule is the presence of phase-angles. In the case of dichotomous observables of the von Neumann-Lüders type the phase angles ϕ_j can be expressed in terms of probabilities. However, in a general case such a straightforward probability interpretation has not yet been found. Pragmatically we can treat these phases as additional parameters of the nonclassical probability inference scheme determined by the quantum formalism. Of course, this is just an operational solution of the problem: the formalism tells us that the quantum probability update has to be of such form (but it is not clear why).

Suppose that now A is a SIC-POVM. Then a-measurements can determine the state ρ precisely. And the first measurement can be simply treated as obtaining prior (with respect to the b-measurement) information about the state of the system. The latter scheme, i.e., with A which is a SIC-POVM, matches QBist treatment of QM as the probability update theory. Of course, QBists treat probability subjectively and by the Växjö interpretation probabilities are treated statistically. However, this is not the only difference. The main difference is that QBists when considering possible modifications of FTP did not take the simplest way, i.e., incorporate an additive perturbation, instead they selected Fuchs-Schack version of FTP, see (6.4). We also point out that in the Växjö approach there is no need to consider the complete state determination, i.e., there is no reason to take A as a SIC-POVM.

Note that the Växjö scheme of probability update based on quantum FTP, see (6.8) and (6.9), can also be used for subjective probabilities. In this way one obtains a new version of QBism - the same private user's interpretation of probability, but another rule for probability update. This version of QBism, so to say QBism-Växjö, can be useful in applications outside of physics (Chapter 9), especially in the process of decision making: decision maker treats probabilities subjectively. The original QBism, i.e., with the update rule (6.4), is also a promising interpretation for such quantum-like modeling.

At the same time, I think that in physics one has to keep to the statistical interpretation of probability which is basic for classical statistical physics. Thus probabilities in (6.8) and (6.9) have to be treated statistically.

We remark that typically the classical Bayesian update of probabilities is presented in the subjective probability framework. (And this was the original approach of Bayes.) However, in many applications, especially in engineering, physics, chemistry, biology, probabilities involved in the Bayesian scheme are interpreted statistically, see Chapter 1, section 1.12, on the views of Gnedenko on this problem.

6.10   Projection Postulate: von Neumann and Lüders Versions

We now discuss interpretational issues related to the von Neumann-Lüders measurements, the formalism presented in sections 4.2.1 and 4.2.2. In Chapter 4 we did not want to mix the presentation of the mathematical structure of QM with interpretational issues; that is why we postponed this discussion.

The projection-form of the feedback reaction, see (4.14), (4.15), to the measurement (resulting in a particular outcome) was postulated by von Neumann [250]. It is well known as the *projection postulate* of QM (the state reduction postulate or the state collapse postulate).

It is less known (in fact, practically unknown) that von Neumann [250] sharply distinguished the cases of observables with non-degenerate spectrum (i.e., all (P_{a_i}) in the spectral decomposition of A, see (4.13), are one dimensional projectors) and degenerate spectrum (i.e., some of (P_{a_i}) are projectors onto multi-dimensional subspaces). For the first case he postulated the aforementioned state-collapse (4.15), but for the second case he pointed out that the measurement feedback can generate state transformations different from the one given by (4.15); in particular, the output state generated from an initial pure state can be a mixed state. Later Lüders [194] extended the von Neumann projection postulate [250] to projectors with degenerate spectra, i.e., in fact, he reduced the class of possible state transformations (quantum operations). This simplification was convenient in theoretical studies and the projection postulate was widely treated as applicable generally, i.e., even to observables with degenerate spectra. The name of Lüders was washed out from the majority of foundational works and nowadays the projection postulate is typically known as the von Neumann projection postulate, although the name of von Neumman should be associated with this postulate only for quantum observables with non-degenerate spectrum, see [157] - [159] for more details.

From first days of QM, physicists widely used the projection postulate without distinguishing the cases of non-degenerate and degenerate spectra. In particular, in their seminal paper [85] Einstein, Podolsky, and Rosen (EPR) apply the projection postulate for local observables on compound systems, but such observables always have degenerate spectra. Thus it is not clear whether the EPR-considerations were approved by von Neumann. In fact, von Neumann considered measurements on compound systems in entangled states in his book [250] (without using the terminology "entangled"). In contrast to the EPR-conclusion [85], von Neumann did not come to the conclusion that QM is either incomplete or nonlocal. For him, as well as for Bohr [38] - [40], it is local and complete. This difference in conclusions is, in fact, a consequence of different viewpoints on applicability of the projection postulate for systems with degenerate spectra. By following von Neumann and saying that, for such observables, the measurement feedback can transfer a pure state into a mixed state we can resolve the "EPR-paradox", see [157] - [159] for details.[20] We note that realization of quantum teleportation scheme is also fundamentally based on the Lüders form of the projection postulate [160]; it seems that, for von Neumann, the applicability of this scheme would be questionable. Moreover, even realization of the quantum computing scheme can be realized only in the Lüders, but not von Neumann approach [160].

Now we turn to a more detailed analysis of von Neumann's sharp dis-

---

[20]Historically it is interesting that von Neumann had never mentioned the EPR-paper (may be because he lost interest in QM?) and Einstein, Podolsky, and Rosen had never appealed to von Neumann's theory of measurement on compound systems.

tinguishing between two types of observables, with non-degenerate and de-generate spectra. He reasoned in the following way. If measurement of some observable A produces the value a_i and if this value is repeated with probability one in the repeated measurement of A, it is natural to assume that after the first measurement the initial pure state ψ was transformed into some vector ψ_{a_i} belonging to the eigenspace H_{a_i} corresponding to the eigenvalue a_i (since only states from this subspace produce the result a_i with probability one). However, there is no reason (from von Neumann's viewpoint) to expect that this output state ψ_{a_i} would coincide precisely with the orthogonal projection on H_{a_i}. This is just one of the vectors be-longing to H_{a_i}. If, for another system prepared in the same initial state ψ, the result of A-measurement is also a_i, then by von Neumann's logic the output state ψ'_{a_i} has to be in H_{a_i}. But it can be different from ψ_{a_i}. Thus in general, for measurements performed for an ensemble of systems in the same state ψ, the output quantum state can be a mixture of states belonging to H_{a_i}, i.e., given by a density operator.

Depending on the interpretations of the quantum state, cardinally dif-ferent interpretations are assigned to the projection postulate.

For those using the realist statistical (ensemble) interpretation of a quantum state, in the spirit of Einstein-Ballentine, this postulate is sim-ply about update of probability distributions (for all possible observables) for an ensemble of identical quantum system prepared for measurements. This is a formal operational rule for finding probability distributions of the output ensemble. Of course, there are no doubts that more general update rules can arise. Similar viewpoint on the projection postulate is charac-teristic for the information interpretations of the quantum state: this is a specific rule for information/probability update. Collapse has no physical meaning, it takes place in the information space. Thus the views of users of the statistical and information interpretations on collapse of a quantum state practically coincide. And this is really surprising. We also remark that L. de Broglie who was the farther of the *double solution* interpreta-tion of QM (which is realist by its nature) also considered collapse of wave function as happening in the information space [67].

For adherents of the orthodox Copenhagen interpretation, the wave function/quantum state is a physical state of an individual quantum system. Therefore quantum operations, measurement feedback state-transformations, take place in physical space by making state collapse one of the most mysterious and intriguing features of quantum theory. Moreover, the state collapse can happen spontaneously, i.e., without any relation to measurement. Considerable experimental efforts have been spent to verify various models of spontaneous collapse.

In general the projection postulate plays a special role in axiomatics of QM. Although majority of the quantum community uses it routinely and operates with the notion of collapse of the wave function, a part of this community handles this postulate with suspicion. This postulate remains controversial, see Auyang [23] p. 23, for the detailed discussion. In particular, Beltrametti and Cassinelli (who are among the world leading experts in quantum logic and foundations) remarked [32] that "it does not have the status of postulates of quantum theory, necessary for its internal coherence."

L. E. Ballentine (who is also one of the world leading experts in quantum foundations, one of the creators of the "statistical interpretation of QM") pointed out [27] that this postulates leads to wrong conclusions. Even if a quantum system through interaction with a measurement device triggers it to produce one fixed eigenvalue, in general the state of this system does not collapse. As an example of inapplicability of the projection postulate, he considered the track left by a quantum (charged) particle in a cloud chamber. Typically the state of the incoming particle is given by a momentum amplitude. The particle ionizes some fist atom it happens to meet in the chamber and this ionization event is considered as position measurement. By the projection postulate the particle state should collapse to the corresponding eigenstate of the position operator, but the latter is a spherical wave which spreads out uniformly in all directions. Hence, it would be impossible for this particle to ionize subsequent atoms to form a track which indicates the direction of original momentum, which is allegedly destroyed by the first ionization event.

The discussion above can be used as an argument for considering quantum instruments of non-projection type, see section 4.5.

Chapter 7

Randomness: Quantum Versus Classical

7.1 Irreducible Quantum Randomness

As was discussed in section 6.4, von Neumann [250] pointed out that quantum randomness is individual, e.g., even an individual electron is intrinsically random, while classical randomness is related to variation of states in an ensemble. In particular, von Neumann remarked [250], pp. 301-302, that, for measurement of some quantity R for an ensemble of systems (of any origin),

"it is not surprising that R does not have a sharp value ..., and that a positive dispersion exists. However, two different reasons for this behavior a priori conceivable:

1. The individual systems $S_1, ..., S_N$ of our ensemble can be in different states, so that the ensemble $[S_1, ..., S_N]$ is defined by their relative frequencies. The fact that we do not obtain sharp values for the physical quantities in this case is caused by our lack of information: we do not know in which state we are measuring, and therefore we cannot predict the results.

2. All individual systems $S_1, ..., S_N$ are in the same state, but the laws of nature are not causal. Then in the cause of the dispersion is not our lack of information, but is nature itself, which has disregarded the principle of sufficient cause."

Thus, for him, quantum randomness is statistical exhibition of violation of causality, violation of the *principle of sufficient cause*. We compare this kind of randomness with classical interpretations of randomness, see Chapter 2:

(1) unpredictability (von Mises),
(2) complexity-incompressibility (Kolmogorov, Solomonoff, Chaitin),
(3) typicality (Martin-Löf).

We start with Kolmogorov algorithmic complexity. One can argue that violation of the principle of sufficient cause has to imply impossibility of nontrivial compression of information in a string of bits produced by measurements of a quantum observable A for an ensemble of systems prepared in the identical state ψ, that is, impossibility to wrtie a program which would produce this sequence and be essentially shorter than the sequence, see Chapter 2 for the definition. Really, it is difficult to imagine how any type of algorithmic process (different from simple output of the result of measurement) can be represented mathematically in the absence of the classical state representation (which would encode possible outputs of measurement). However, it is not easy to formalize this kind of reasoning. Probably, here we suffer from the lack of imagination, see also citation of Bell below.

Thus if the principle of sufficient cause is really violated for quantum systems and one can really connect its violation with impossibility to shorten representation of the sequence of results of measurement, then the output sequences of quantum measurements must have high Kolmogorov complexity and quantum measurements can be used for creation of random numbers (in the sense of Kolmogorov's complexity approach). This reasoning provides, in fact, the foundational (philosophical) basis for the project on *quantum random generators* [217].

Thus justification of proper functioning of quantum random generators can be done only from physical principles. Surprisingly mathematics still plays a crucial role, since it is heavily explored in so-called no-go theorems saying that it is impossible to introduce hidden variables, parameters which provide a finer description of the state of a quantum system than given by its quantum state. Violation of the principle of sufficient cause is incompatible with the existence of subquantum models with hidden variables. Roughly speaking, one cannot start to sell quantum random generators before a loophole free test rejecting existence of hidden variables is successfully performed.[1]

Von Neumann understood very well the role of no-go theorems in justification of his thesis about violation of causality by quantum systems and he formulated and presented [250] a sketch of proof of the first no-go theorem, nowadays known as the *von Neumann theorem*. However, his theorem was

---

[1]Recently experimenters performed a few tests, e.g., [109], [102] which pretend to be loophole free. However, it is too early to declare that the problem of loopholes was completely solved. The experimental data should pass independent statistical analysis, cf. [260], [169].

criticized, e.g., by Bell [29], [31] (see also introduction to Chapter 8) and Ballentine [25], [26] as based on unphysical assumptions about the rules for coupling an imaginable classical hidden variable model of quantum phenomena and the genuine quantum model. This theorem is considered as having no physical impact.

Now the most fashionable no-go theorem is due to Bell [31], see Chapter 8 for the detailed presentation. However, as well as the von Neumann theorem, it is based on concrete rules coupling imaginable classical hidden variable model and the genuine quantum model [161]. Adequateness of these rules to real physical situations can also be criticized [67], [2] - [4], [133], [156], [161], [188] - [191], [111] - [113], [72] - [74], [203], [185], [186], [206] .... This is, in fact, the main problem of all no-go theorems [161], attempts to reject all possibly imaginable models with hidden variables and imaginable rules coupling them with the quantum model (see also introduction to Chapter 8). Even Bell pointed out [30] "that what is proved, by impossibility proofs, is lack of imagination."

Another problem is that we still have no no-go theorem which validity was completely certified experimentally (with recognized statistical justification), see [101], [59], [109]. [102] for recent experimental tests claiming closing all basic loopholes; see also, e.g., [260], [169] for recent contribution to statistical analysis of Bell's tests.

Let us assume (for a moment) that Bell's theorem really can be considered as describing physically reasonable coupling between the most general model with hidden variables and QM (as the majority of the quantum community believes). There is still one fundamental problem preventing justification of von Neumann's statement about violation of the principle of sufficient cause. Bell's theorem rejects only local hidden variable models, see Chapter 8 for details, i.e., models preventing faster than light communications. Thus von Neumann was right and the quantum random generators really produce random sequences (at least in the framework of Kolmogorov's algorithmic complexity) only if nature were not too exotic, i.e., superluminal communications were impossible.

As we know (Chapter 2), the Kolmogorov and Martin-Löf approaches to randomness lead to the same class of random sequences. Therefore one can proceed formally and say that the above reasoning that quantum randomness implies Kolmogorov's randomness also leads to the implication: quantum randomness↦ Martin-Löf randomness. Thus a sequence produced by a quantum random generator has to pass the universal Martin-Löf test and, in particular, any finite block of algorithmic tests, e.g., the NIST test. Such type of reasoning is very popular in the community working with quantum random generators. It seems that one needs not to check whether the output of a quantum random generator would pass, e.g., the NIST test. There are a few objections to such a viewpoint. The first one is based on recognition that each quantum experiment depends on numerous "technicalities" modifying the output. That is, the actual output may essentially differ from output expected from the theoretical analysis of the experimental design. Thus, in any event, the NIST test is needed to certify a quantum random generator. Another objection is of a fundamental character – the absence of experimentally justified no-go theorem.

Now we turn to the notion of randomness as unpredictability, a la von Mises. We repeat that von Mises' principle of randomness (Chapter 1) can be treated as *the law of excluded gambling strategy.* However, such a strategy definitely does not exist if the principle of sufficient cause is violated. Hence, under the latter assumption outputs of quantum measurements can be considered as random, from the von Mises viewpoint, i.e., as unpredictable.

We point out that violation of the principle of sufficient cause is state dependent. If the state ψ of a system is one of the eigenstates of the operator A representing a quantum observable, then we can predict the result of measurement with probability one. Thus this principle has to be used with caution.

All previous considerations were devoted to matching of the notion of quantum randomness to the standard notions elaborated in mathematics. As we have seen, by assuming that a sequence is produced by a quantum source of randomness one can be sure that it is random in the classical sense. Thus to be random in the classical sense is a necessary condition of quantum randomness. Is it sufficient? The canonical answer is "no". It is typically claimed that only quantum randomness is genuine randomness, see section 7.3 for further discussion.

7.2   Lawless Universe? Digital Philosophy?

Where did complexity of the Universe come from? This is one of the most fundamental problems of modern science. One of the first scientists who took this problem seriously was Leibniz, see Chaitin [55] for the excellent presentation of Leibniz's views on this problem.[2] And to explain the origin of Universe's complexity Leibniz "simply" appealed to God.

Leibniz was interested in distinguishing lawful and lawless experimental data. He presented a beautiful example illustrating this problem. He proposed us to scatter points at random on a sheet of paper, closing the eyes and stabbing at the paper with a pen many times, say a few hundreds. The output will be a randomly looking pattern on the sheet. However, Leibniz pointed out that even for this data one can easily find a mathematical law, in fact, a polynomial curve, that passes through all these randomly chosen points. To show this he used the concrete application of Lagrangian interpolation procedure.

For reader's convenience, we recall that the Lagrange interpolating polynomial is the polynomial $P(x)$ of degree $\leq (n-1)$ that passes through the n points $(x_1, y_1 = f(x_1)), (x_2, y_2 = f(x_2)), ..., (x_n, y_n = f(x_n))$, and is given by

$$P(x) = \frac{(x - x_2)(x - x_3)...(x - x_n)}{(x_1 - x_2)(x_1 - x_3)...(x_1 - x_n)} y_1 \tag{7.1}$$

$$+ \frac{(x - x_1)(x - x_3)...(x - x_n)}{(x_2 - x_1)(x_2 - x_3)...(x_2 - x_n)} y_2 + ...$$

$$+ \frac{(x - x_1)(x - x_2)...(x - x_{(n-1)})}{(x_n - x_1)(x_n - x_2)...(x_n - x_{n-1})} y_n. \tag{7.2}$$

Thus, in spite of the fact that the generation of the aforementioned pattern satisfies the heuristic criteria of randomness as unpredictability, we cannot say that the output pattern is lawless. It seems that Leibniz was the first who framed the problem of distinguishing lawfulness and lawlessness correctly: not as distinguishing between lawful and totally lawless processes, but between processes having different degree of complexity. We see that the complexity of the mathematical law (7.2) rapidly increases with the increase of the size of the pattern. As was emphasized by Chaitin [55], this was the first step towards the modern theory of complexity and randomness. Following Leibniz, we can say that the basic task of a scientist is not just to find mathematical laws describing natural (or mental) phenomena, but the simple laws, laws of low complexity, which at the same time

---

[2]This book is the apotheosis of the algorithmic approach, not only to randomness and complexity, but to science in general. (Of course, it is surprising that in this book Kolmogorov was not mentioned at all!) By presenting digital philosophy in its extreme form Chaitin's book serves perfectly for our purpose: criticism of this philosophy. Therefore this book will be often cited in coming discussions.

produce sufficiently rich patterns of data (to be of some interest for scientists). Again from the above toy example, one can see that the "majority of mathematical laws" are complex, the appearance of a simple law with rich output is merely an exception. And, for Leibniz, such exceptionally simple and fruitful laws could appear only in accordance with God's plan: "God has chosen that which is the most perfect, that is to say, in which at the same time the hypotheses are as simple as possible, and phenomena are as rich as possible." Here Leibniz was cited again by following Chaitin [55] who used this citation for the following passage:

"The complexity of the Universe is combined from the complexity of laws for expressing of natural phenomena and the complexity of initial and boundary conditions. If initially the Universe was described by simple initial-boundary conditions (as by the Big Bang scenario), then the complexity of the Universe is due to complexity of laws for expression of natural phenomena. Thus the complexity of the Universe can be identified with complexity of its laws and the latter has to be measured as the algorithmic complexity".[3]

However this way of thinking cannot explain where this complexity of laws came from. Although Leibniz's views are very sympathetic for Chaitin, he is not consistent enough to appeal to God. Instead of such an appeal to God's plan of creation of the perfect world, he referred to *quantum randomness* as generating patterns which cannot be described by simple laws. However, in the purely classical considerations his reference to quantum is really illogical. How can one refer to quantum randomness if the usual classical coin tossing generates a pattern which is algorithmically so complex as a pattern produced by a quantum random generator? We cite book [55] again, p. 119:

"This idea of an infinite series of independent tosses of a fair coin may sound like a simple idea, a toy physical model, but it is a serious challenge, indeed a horrible nightmare, for any attempt to formulate a rational world view! Because each outcome is in fact that is true for **no reason**, that is true only by accident!"

Chaitin can really be considered as one of the fathers of *digital philosophy* which, in particular, led to *digital physics* culminating in Wheeler's statement [257]:

"It from bit. Otherwise put, every 'it', every particle, every field of force, even the space-time continuum itself derives its function, its meaning, its very existence entirely - even if in some contexts indirectly from the apparatus-elicited

---

[3]My comment: Of course, it is questionable whether a singularity can be treated as a simple initial-boundary condition.

answers to yes-or-no questions, binary choices, bits. 'It from bit' symbolizes the idea that every item of the physical world has at bottom - a very deep bottom, in most instances - an immaterial source and explanation; that which we call reality arises in the last analysis from the posing of yes-no questions and the registering of equipment-evoked responses; in short, that all things physical are information-theoretic in origin and that this is a participatory universe."

We also cite Chalmers [56] comment on this statement:

"Wheeler (1990) has suggested that information is fundamental to the physics of the universe. According to this 'it from bit' doctrine, the laws of physics can be cast in terms of information, postulating different states that give rise to different effects without actually saying what those states are. It is only their position in an information space that counts."

In this book we are mainly interested in quantum theoretical version of informational physics, as Zeilinger-Brukner informational interpretation and QBism of Fuchs et al. (see Chapter 6) and D'Ariano et al. operational-information derivation of QM. These approaches, while being a part of information physics, do not match the digital philosophy. Here information is considered as a primary physical quantity which cannot be defined in terms of other more fundamental variables. Opposite to Chaitin and Wheeler, they put the transcendental content in the notion of information which matches perfectly with the transcendental content of a quantum state.

We also mention the information viewpoint on Bohmian mechanics based on the *active information* interpretation of the quantum potential. This interpretation was elaborated by Bohm and Hiley [36]. It is amazing that even an ontological model of quantum phenomena, Bohmian mechanics, naturally generates the purely information interpretation of its basic entity, the quantum potential.[4]

7.3 Unpredictability and Indeterminism

Unpredictability is very often coupled with indeterminism. The latter is impossibility to describe generation of data by a dynamical map:

$$y = U(x_0), \tag{7.3}$$

where U is a map from the input x_0 to the output x.

---

[4]The active information interpretation opened the door for applications of the formalism of Bohmian mechanics outside physics, in particular, in mathematical modeling of quantum-like cognition [36], later this formalism was explored in behavioral finances [133], [156], [106].

In the simplified picture of random processes determinism implies predictability, so no randomness. To be unpredictable a process has to be
indeterministic. However, this picture does not match the real situation.

Consider the basic example of a collective, Mises random sequence: an
infinite series of independent tosses of a fair coin. It is also random from
the viewpoint of Kolmogorov, i.e., in the framework of algorithmic complexity. Hence, it is Martin-Löf random. However, a coin is a classical
mechanical system, and its motion is described by Newtonian mechanics, [252], [75], [204], [239]. Hence, one can construct the corresponding
dynamical map (7.3). Thus if we know the initial condition, we can predict the outcome of a coin toss. The process of generation of this (Mises-
Kolmogorov-Martin-Löf) random sequence is totally deterministic. Its unpredictability is just a matter of imprecision in determination of initial conditions. Nowadays this trade between (un)predictability and (im)precision
in determination of x_0 can be numerically modeled [239]. The latter paper
contains a detailed mechanical model of the coin tossing dynamics. The
results of the corresponding numerical simulation were presented graphi-
cally. It was shown that if the imprecision in selection of x_0 is less than ϵ,
where ϵ depends on parameters of the model (see [239] for details), then one
can predict the outcome of each coin toss. However, if one can determine
x_0 with accuracy only up to a ball of some radius larger than this ϵ, then
the outcome cannot be determined in advance. Thus the story about coin
tossing is a story about the precision of determination of initial conditions.
Since this is one of the basic examples of the Mises-Kolmogorov-Martin-
Löf random sequences, we conclude that the modern mathematical theory
of randomness does not contradict determinism in sequence generation.
That is why Chaitin [55] has to refer to quantum randomness to emphasize the lawlessness dimension of randomness. To be more precise, we have
to speak about the complexity dimension. However, the example of coin
tossing shows that there is nothing about complexity of physical laws. The
dynamical equations [239] are simple Newtonian equations. At the same
time Kolmogorov complexity of the output is very high. What does this
mean? Simply that *Kolmogorov complexity is not an adequate measure of
complexity of physical laws behind generation of sequences of outputs.*

What is the main problem in matching the Kolmogorov approach with
physics? This is consideration of solely algorithmically representable laws.
The algorithmic-computability approach well serves the purposes of computer science and artificial intelligence, but not physics. All basic physical
models contain some *transcendental* element. For example, Newtonian me-

chanics is based on *real numbers*. We remind a few measure-theoretic facts about reals.

- Consider the segment $[0,1]$ and probability p_L given by the linear Lebesgue measure; here $p_L([a,b]) = b - a$. Then probability that a randomly selected number from $[0,1]$ is rational equals to zero, the same is valid for algebraic numbers (solutions of algebraic equations). Thus probability to get a transcendental number is one.
- One can introduce the notion of a *computable real number*. Real numbers with probability one are noncomputable.

The classical model of natural phenomena is fundamentally noncomputable. The main problem of digital philosophy and digital physics is that they try to identify the human brain with computer. The latter definitely can operate only with computable quantities, but the former can easily make transcendental steps in reasoning, see R. Penrose [216] for detailed presentation of this viewpoint.

Therefore the following statement of Chaitin, *"the manifest of computability"*, is not about science done by humans, but science done by computers or other artificial intellectual systems, so see [55], p. 64:

"I think of a scientific theory as a binary computer program for calculating the observations, which are also written in binary. And you have a law of nature if there is compression, if the experimental data is compressed in a computer program that has a smaller number of bits than are in the data that it explains.
...

But if the experimental data cannot be compressed, if the smallest program for calculating it is just as large as it is ..., then the data is lawless, unstructured, patternless, not amenable to scientific study, incompressible. In a word, random, irreducible!"

Now we turn to quantum randomness. As was pointed out, the mathematical theory of randomness cannot distinguish "classical and quantum randomness", random sequences generated by coin tossing and by quantum random generators. They are equally algorithmically complex (Kolmogorov) and typical (Martin-Löf).[5] How can one try to formalize the notion of quantum random sequence? Combing the viewpoints of von Neumann and Kolmogorov-Martin-Löf (and Church, Solomonoff, Chaitin, Schnorr), we can say that this is a Kolmogorov-Martin-Löf random sequence

---

[5]We remind that "a good theory of randomness as unpredictability", a la von Mises, has not yet been created. Its development culminated in Wald's theorem, Theorem 2, Chapter 2, section 2.1.1. The next step, to the Church-Wald collectives, might be a step in the wrong direction.

such that it is impossible to present a causal model of its generation. (A larger class of quantum randomness one gets by considering a deterministic dynamical system, instead of a general causal model.)

However, it seems to be impossible to prove the impossibility of causal generation for a concrete sequence. In spite of huge activity in generation of various no-go theorems, we still do not have an adequate no-go theorem for one output experiment with quantum systems. The famous Bell theorem is about impossibility of combination of a few causally generated outputs (if we ignore the issue of nonlocality). This theorem cannot exclude the possibility that each of them can be causally generated. It even cannot exclude the possibility of deterministic generation of each of these outputs. An adequate no-go theorem might be the original von Neumann theorem [250], see also section 7.1. However, nowadays it is commonly considered as inadequate to the real quantum mechanical situation.

Is it possible that sooner or later it will become evident that genuine quantum randomness does not exist? Can it be just randomness due to imprecision in selection of initial and boundary conditions? A randomly selected representative of the quantum community would answer: "no!!!" However, this "no" is merely an expression of the degree of the personal belief and not estimation of the real experimental situation.

As was pointed out by H. Rauch, we do not know initial and boundary conditions in, e.g., neutron interferometry experiments. Therefore it is not surprising (from the very classical viewpoint) that we are not able to predict experiment's output. One cannot exclude (at least in the case of experimenting with neutrons which are heavy quantum objects) that it may be possible to determine initial conditions more precisely and to get more information than given by the operational quantum formalism.[6] Of course, A. Zeilinger, the former PhD-student of H. Rauch, would strongly protest against such a viewpoint by arguing that randomness of quantum behavior cannot be reduced to selection of initial and boundary conditions. This huge difference in views of a teacher and his former student might be explained (but not uniquely!) by the fact that they work with very different representatives of the quantum world, neutrons and photons, respectively.

Finally, we remark that one has to distinguish quantum randomness from quantum probability. It seems that these notions are often identified

---

[6]Helmut Rauch is one of the world's top experimenters in the domain of neutron interferometry. In particular, he built the first neutron interferometer at the Atom Institute of the Austrian Academy of Science. During the Växjö series of conferences on quantum foundations, 2000-2015, he expressed this viewpoint on many occasions.

(may be unconsciously). Then the nonclassical structure of quantum probability is treated as the argument in favor of nonclassicality of quantum randomness. For example, we can point to intensive studies justifying peculiarities of quantum randomness as compared to classical randomness by using the Bell no-go theorem (see Chapter 8). Here nonclassical probability structure of the Bell test is treated as leading to generation of nonclassically random sequences.

Chapter 8

Probabilistic Structure of Bell's Argument

By starting to study some scientific theory it is very useful to know philosophic views of its creators. Regarding Bell's inequality project we have to be aware that J. Bell was one of those using the realist interpretation of QM, as well as, e.g. Einstein, Podolsky, and Rosen, De Broglie, early Schrödinger, ..., D. Bohm, L. Ballentine, the author of this book,... . In contrast to Einstein, Bell (following Bohm) did not consider nonlocality as "an absurd alternative to incompleteness of QM". As well as for Bohm, the universe was the indivisible entity, cf. with the book of D. Bohm and B. Hiley [36] about "undivided universe." This philosophy of wholeness played the determining role in Bell's research on quantum foundations. Roughly speaking his main aim (at least at the beginning) was to justify Bohmian mechanics.[1] Thus by studying Bell's argument it is important to be aware that *Bell was Bohmian.*

In particular, one of his first steps towards research on quantum foundations was the attack on the von Neumann no-go theorem [250]. Nowadays this first no-go statement about hidden variables is practically forgotten. My private experience shows that the young generation working in quantum physics, even in quantum information and quantum foundations did not hear about it. Therefore I shall say a few words about it, although this is not a topic of this book. This statement can be considered as a formal

---

[1]This is a good place to remark that the common identification of the views and approaches of De Broglie and Bohm, e.g., in the form of the statement that "Bohm created a consistent version of De Broglie's double solution theory", is not justified. In context of the present discussion it is especially important to point out that De Broglie did not treat his double solution theory as nonlocal. He separated sharply physical nonlocality and "mathematical nonlocality" present in differential equations of his theory. The latter was a special form of representation of quantum correlations and not at all action at the distance, read, e.g., De Broglie's book [67], including the appendix.

mathematical reply to the EPR-argument on incompleteness of QM. Von Neumann was sure that he proved that QM is complete and *any naturally formalized* model with hidden variables should contradict to probabilistic predictions of QM. (We remark that the argument of nonlocality was not involved in von Neumann's reasoning.) As was already remarked in this book, the main problem of all no-go statements is that nowadays we do not know either about the mathematical structure of the right subquantum model or about the form of its mapping onto the quantum model. Each author of a no-go statement presents his own views on these issues and then he proves that under such assumptions there is no way to construct a subquantum model. Then others start to criticize these assumptions and this concrete no-go statement. And J. Bell criticized (in quite aggressive manner) the assumptions of the von Neumann no-go theorem. One of such assumptions is that the sum of classical observables ξ and η, classical random variables, has to be mapped onto the sum of the corresponding quantum observables represented by the Hermitian operators a and b. For Bell, this attack on von Neumann's reasoning was an important step to motivate his further studies directed to justification of Bohmian nonlocality. Otherwise, if QM were "simply incomplete" (as was "proved" by von Neumann), it would be unnecessary to attract the nonlocality argument. This is a good place to say that, for Copenhagenists, Bohr, Heisenberg, Pauli, von Neumann, Dirac, Bell's reasoning including his famous inequality would be totally meaningless, because they stayed on the position that QM is complete (for Bohr and Heisenberg, this was a consequence of Heisenberg's uncertainty principle).

The previous discussion can be summarized by saying that, for Bell, from the very beginning nonlocality and realism were the main issues. This explains why, in contrast to Feynman [88], section 5.1, Bell did not pay so much attention to the probability structure of his argument. In particular, he never discussed Feynman's viewpoint on nonclassicality of quantum probability and, hence, correlations. For Bell, they are classical, but nonlocal.

Feynman's statement about nonclassicality of the probabilistic structure of the two-slit experimental test stimulated me to perform analysis of the probabilistic structure of experimental tests on violation of the Bell-type inequalities. The similarity of these tests (two-slit and Bell-type) is evident. Both are based on combination of a few incompatible experimental contexts. In the two-slit experiment these are three contexts corresponding to three possible combinations of two slits opening-closing; in, e.g., the test on violation of CHSH inequality these are four contexts corresponding to combinations of two angles of two Polarization beam splitters (PBSs). One can guess that, similarly to the two-slit experiment, contextuality is responsible for nonclassicality of the CHSH-correlations. (Once again: we use the notion of contextuality in more general sense than it is nowadays common in quantum foundations; our contextuality is of the Bohr type - dependence on the whole experimental arrangement, not just on measurement of a compatible observable.)

By using contextuality it is important to proceed in the formal mathematical framework and we presented the mathematical contextual probabilistic model in section 5.5. In this contextual probabilistic framework we sharply distinguished two types of compatibly: unconditional and conditional. *Unconditional compatibility* is equivalent to noncontextuality (the latter is treated in our general framework). By ignoring such metaphysical issues as non/realism, non/locality we can say that the Bell statement is about impossibility to treat observables involved in the Bell-type tests as unconditionally compatible, i.e., without taking into account context selection. However, the intrinsically contextual structure of the Bell-type tests makes this evident from the very beginning, at least if one has Feynman's image of the probabilistic structure of the two-slit experiment. Of course, one should check not the assumption of unconditional compatibility, as Bell proposed to do, but the assumption of conditional compatibility. If the latter were violated, then, of course, we might feel that something mystical happens. However, again from the very beginning one can hope to construct a mathematical representation of the Bell-test observables as conditionally compatible. Such model was constructed by the author in [168] (see also for the earlier version [24]: here I used the frequency model of von Mises). And the main aim of this chapter is presentation of this model. (See also Appendix B - comparing embedding of quantum probability model into the Kolmogorov model with embedding of non-Euclidean geometry models into the Euclidean one.)

We also remark that in classical probabilistic modeling of quantum correlations contextuality was (implicitly) explored by many authors, see, e.g., Kupczynski [188] - [192], Hess and Philipp [111] - [113], De Raedt et al. [72] - [74]. This contextual dimension is typically not recognized in critical attacks to these papers which is not surprising, since the authors of aforementioned works do not emphasize contextuality of their models - using of conditioning of classical probabilities (may be some of them neither recognize this). I explain once again my own position [133] - [135], [155] - [160].

Bell's argument [31], as was formulated, is, of course, correct for his classical model of quantum measurements. However, this model is highly idealized: experimental arrangement is not taken into account. (Thus, e.g., Bohr, Pauli, Heisenberg would reject it immediately without debates on violation of Bell's inequality.) Any attempt to take experimental contexts into account immediately leads to probability conditioning constrains. Classical models of measurement taking into account such constraints can reproduce predictions of QM, including correlations for entangled systems. Thus my main critique is not against so to say "Bell's theorem", but againts his model of measurement. (It is clear that it has its roots in von Neumann's argumentation against hidden variables [250]. Bell criticized sharply von Neumann's argumentation, but at the same time he borrowed von Neumann's idealized model.) The same contextual (conditional probability) approach resolves other "paradoxes" of quantum-classical interrelation, e.g., the GHZ-paradox [140].

8.1 CHSH-inequality in Kolmogorov Probability Theory

In Chapter 1 we derived one of the Bell-type inequalities, the Wigner-Bell inequality. However, in real experiments another version of the Bell-type inequalities plays the important role, namely, the CHSH inequality, where CHSH stands for John Clauser, Michael Horne, Abner Shimony, and Richard Holt. We present it as a theorem of Kolmogorov probability theory.

Let $\mathcal{P} = (\Omega, \mathcal{F}, p)$ be a Kolmogorov probability space. For two random variables A and B, we set

$$< A, B >= E(AB) = \int_\Omega A(\omega)B(\omega)dp(\omega).$$

Theorem 1. (CHSH-inequality) *Let $A^{(i)}, B^{(j)}, i, j = 1, 2$, be random variables with values in [-1,1]. Then the corresponding combination of correlations*

$$S =< A^{(1)}, B^{(1)} > + < A^{(1)}, B^{(2)} > + < A^{(2)}, B^{(1)} > - < A^{(2)}, B^{(2)} >, \tag{8.1}$$

satisfies the CHSH-inequality:

$$|S| \leq 2. \tag{8.2}$$

Proof. It is easy to show that for any quadruple of random variables valued in [-1,1] the following inequality holds:

$$2 \leq A^{(1)}(\omega)B^{(1)}(\omega) + A^{(1)}(\omega)B^{(2)}(\omega) + A^{(2)}(\omega)B^{(1)}(\omega) - A^{(2)}(\omega)B^{(2)}(\omega) \leq 2.$$

By integrating this inequality with respect to the probability measure p we obtain (8.2).

In quantum foundations the main issue is that, for some quantum states and observables, correlations can violate the CHSH-inequality. What does it mean from the viewpoint of probability theory? As was pointed out in Chapter 5, this means that contexts used for calculations of these correlations are probabilistically incompatible in the sense of Definition 2, section 5.5.3, Chapter 5, i.e., they are not unconditionally probabilistically compatible. It is impossible to construct Kolmogorov probability space $\mathcal{P} = (\Omega, \mathcal{F}, P)$ such that all observables $A^{(i)}, B^{(j)}$ were represented as random variables belonging to $R(\mathcal{P})$.

However, in quantum foundational studies a violation of the CHSH inequality by some quantum correlations is not interpreted in such purely probabilistic manner. Instead of speaking about non-Kolmogorovness of multi-contextual probabilistic data, experts in quantum foundations operate with the notions of locality and realism. If we translate these notions to the language of probability theory, we get precisely Kolmogorovness (probabilistic unconditional compatibility).

As was emphasized in Chapter 5, and the introduction to the present chapter, the experimental test under consideration is intrinsically multi-contextual. Therefore it would be surprising if this multi-contextual structure were unconditionally representable. One has to look for a conditional probabilistically compatible structure. We shall present it in the following sections.

8.2 Bell-test: Conditional Compatibility of Observables

Now we present embedding of the probabilities (and correlations) for joint measurements of polarizations for pairs of photons given by QM and violating the CHSH-inequality (8.2) into a Kolmogorov probability space.

To verify an inequality of this type, one should put in it statistical data collected for *four pairs of settings* of polarization beam splitters (PBSs):

$$\theta_{11} = (\theta_1, \theta_1'), \theta_{12} = (\theta_1, \theta_2'), \theta_{21} = (\theta_2, \theta_1'), \theta_{22} = (\theta_2, \theta_2').$$

Here $\theta = \theta_1, \theta_2$ and $\theta' = \theta_1', \theta_2'$ are selections of angles for orientations of respective PBSs. We can speak about experimental contexts $C_{\theta_{ij}}, i, j = 1, 2$. And, as was pointed out, they are not uncoditionally compatible.

The selection of the angle θ_i determines the corresponding polarization observable, $a_{\theta_i} = \pm 1$. There are two detectors coupled to the PBS with the fixed θ-orientation: "up-polarization" detector and "down-polarization" detector. A click of the up-detector assigns to the random variable $a_\theta(\omega)$ the value $+1$ and a click of the down-detector assigns to it the value -1. In the same way selection of the angle θ' determines the corresponding polarization observable, $b_{\theta'_i} = \pm 1$.

We repeat once again that the CHSH-test consists of four different experiments corresponding to settings θ_{ij}. Our aim is to unify these four experiments into a single experiment with random selection of experimental settings. In principle, such unification is used in modern tests of the CHSH-inequality in which settings are selected with the aid of random generators.

For the illustrative purpose, it is more useful to map this experiment with random switching of orientations of two fixed PBSs onto the experiment in which all settings are unified at the "hardware level", i.e., the experiment with 4 PBSs oriented with the angles θ_1, θ_2 and θ'_1, θ'_2 and each PBS is equipped with its own two detectors, so there are in total 8 detectors.

Such an experimental scheme was used in the pioneer experiment of A. Aspect [17], [18] with one difference: he used single channel PBSs. We, finally present the corresponding citation of Aspect [17]:

"We have done a step towards such an ideal experiment by using the modified scheme shown on Figure 15. In that scheme, each (single-channel) polarizer is replaced by a setup involving a switching device followed by two polarizers in two different orientations: a and a' on side I, b and b' on side II. The optical switch $C1$ is able to rapidly redirect the incident light either to the polarizer in orientation a, or to the polarizer in orientation a'. This setup is thus equivalent to a variable polarizer switched between the two orientations a and a'. A similar setup is implemented on the other side, and is equivalent to a polarizer switched between the two orientations b and b'. In our experiment, the distance L between the two switches was 13 m, and L/c has a value of 43 ns. The switching of the light was effected by home built devices, based on the acousto-optical interaction of the light with an ultrasonic standing wave in water. The incidence angle (Bragg angle) and the acoustic power, were adjusted for a complete switching between the 0th and 1st order of diffraction."

Figure 15 can be found in [17], see also [19]. The only difference of our scheme is that each of the four PBSs has two output channels.

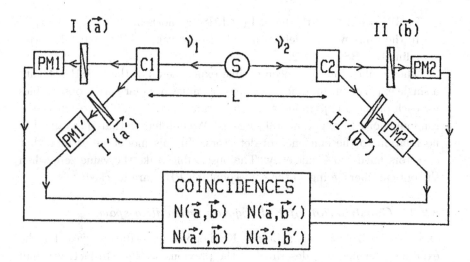

Fig. 8.1 The scheme of the pioneer experiment of A. Aspect with four beam splitters.

8.2.1 *Random choice of settings*

a). There is a source of entangled photons.

b). There are 4 PBSs and corresponding pairs of detectors for each PBS, in total 8 detectors. PBSs are labeled as $i = 1, 2$ (at the left-hand side, LHS) and $j = 1, 2$ (at the right-hand side, RHS).

c). Directly after source there are 2 distribution devices, one at LHS and one at RHS. At each instance of time, $t = 0, \tau, 2\tau, \ldots$ each device opens the port to only one (of two) optical fibers going to the corresponding two PBSs. For simplicity, we suppose that each pair of ports (i, j), $(1, 1), (1, 2), (2, 1), (2, 2)$, can be opened with equal probabilities[2]:

$$\mathbf{P}(i, j) = 1/4.$$

Now we introduce the observables measured in this experiment. They are modifications of the polarization observables $a_{\theta_i}, i = 1, 2$, and $b_{\theta'_j}, j = 1, 2$. We start with the "LHS-observables":

1) $A^{(i)} = \pm 1, i = 1, 2$ if the corresponding (up or down) detector is coupled to ith PBS (at LHS) fires and the i-th channel is open;

---

[2]In general for selection of experimental settings we can use any probability distribution, $\sum_{i=1, j=1}^{2} \mathbf{P}(i, j) = 1$. Under the natural assumption that the distribution devices operate independently, we introduce two probability distributions $P_L(i), i = 1, 2, P_L(1) + P_L(2) = 1$, and $P_R(j), j = 1, 2, P_R(1) + P_R(2) = 1$, and set $\mathbf{P}(i, j) = P_L(i)P_R(j)$.

2) $A^{(i)} = 0$ if the i-th channel (at LHS) is blocked.

In the same way we define the "RHS-observables" $B^{(j)} = 0, \pm 1$, corresponding to PBSs $j = 1, 2$.

Thus unification of 4 incompatible experiments of the CHSH-test into a single experiment modifies the range of values of polarization observables for each of the 4 experiments; the new value, zero, is added to reflect the random choice of experimental settings. We emphasize that this value has no relation to the efficiency of detectors. In this model we assume that detectors have 100% efficiency. The observables take the value zero when the optical fibers going to the corresponding PBSs are blocked.

8.2.2   *Construction of Kolmogorov probability space*

Our aim is to construct a proper Kolmogorov probability space for the experiment which was described in the previous section. In fact, we shall present a general construction for combining probabilities produced by a few experiments which can be incompatible.

In the CHSH-test we operate with probabilities $p_{ij}(\epsilon, \epsilon'), \epsilon, \epsilon' = \pm 1$, – to get the results $a_{\theta_i} = \epsilon, b_{\theta'_j} = \epsilon'$ in the experiment with the fixed pair of orientations (θ_i, θ'_j). From the QM-formalims, we know that, for the singlet state, these EPR-Bohm probabilities are given by expressions:

$$p_{ij}(\epsilon, \epsilon) = \frac{1}{2} \cos^2 \frac{\theta_i - \theta'_j}{2}, p_{ij}(\epsilon, -\epsilon) = \frac{1}{2} \sin^2 \frac{\theta_i - \theta'_j}{2}. \qquad (8.3)$$

However, this special form of probabilities is not important for us. Our construction of unifying Kolmogorov probability space works well for any collection of probabilities

$$p_{ij} > 0 : \sum_{\epsilon, \epsilon'} p_{ij}(\epsilon, \epsilon') = 1. \qquad (8.4)$$

Thus we proceed in this general situation, (8.4). Hence, we consider the experiment described in section 8.2.1 and producing some collection of probabilities (p_{ij}). The special choice of the probabilities (8.3) will be used only to concrete considerations and to violate the CHSH-inequality, see section 8.2.5.

Let us now consider the set of points Ω (the space of "elementary events" in Kolmogorov's terminology):

$$\Omega = \{(\epsilon_1, 0, \epsilon'_1, 0), (\epsilon_1, 0, 0, \epsilon'_2), (0, \epsilon_2, \epsilon'_1, 0), (0, \epsilon_2, 0, \epsilon'_2)\},$$

where $\epsilon = \pm 1, \epsilon' = \pm 1$. These points correspond to the following events: e.g., $(\epsilon_1, 0, \epsilon'_1, 0)$ means: at LHS the PBS with $i = 1$ is coupled and the

PBS with $i = 2$ is uncoupled and the same situation is at RHS: the PBS with $j = 1$ is coupled and the PBS with $j = 2$ is uncoupled; the result of measurement at LHS after passing the PBS with $i = 1$ is given by ϵ_1 and at RHS by ϵ_1'.

We define the following probability measure[3] on Ω

$$\mathbf{P}(\epsilon_1, 0, \epsilon_1', 0) = \frac{1}{4}p_{11}(\epsilon_1, \epsilon_1'), \mathbf{P}(\epsilon_1, 0, 0, \epsilon_2') = \frac{1}{4}p_{12}(\epsilon_1, \epsilon_2')$$

$$\mathbf{P}(0, \epsilon_2, \epsilon_1', 0) = \frac{1}{4}p_{21}(\epsilon_2, \epsilon_1'), \mathbf{P}(0, \epsilon_2, 0, \epsilon_2') = \frac{1}{4}p_{22}(\epsilon_2, \epsilon_2').$$

We now define random variables $A^{(i)}(\omega), B^{(j)}(\omega)$:

$$A^{(1)}(\epsilon_1, 0, \epsilon_1', 0) = A^{(1)}(\epsilon_1, 0, 0, \epsilon_2') = \epsilon_1, \tag{8.5}$$

$$A^{(2)}(0, \epsilon_2, \epsilon_1', 0) = A^{(2)}(0, \epsilon_2, 0, \epsilon_2') = \epsilon_2; \tag{8.6}$$

$$B^{(1)}(\epsilon_1, 0, \epsilon_1', 0) = B^{(1)}(0, \epsilon_2, \epsilon_1', 0) = \epsilon_1', \tag{8.7}$$

$$B^{(2)}(\epsilon_1, 0, 0, \epsilon_2') = B^{(2)}(0, \epsilon_2, 0, \epsilon_2') = \epsilon_2', \tag{8.8}$$

and we put these variables equal to zero in other points.

Remark 1. (Locality of the model) We remark that the values of the LHS-variables $A^{(1)}(\omega), A^{(2)}(\omega)$ depend only the first two coordinates of ω and the values of the RHS-variables $B^{(1)}(\omega), B^{(2)}(\omega)$ depend only on the last two coordinates of ω. Thus the value of $A^{(i)}(\omega)$ does not depend on the values of $B^{(j)}(\omega)$. They neither depend on the action of the RHS distribution device; for the LHS-variable it is not important which port is open or closed by the RHS distribution device. The RHS-variables $B^{(j)}(\omega)$ behave in the same way. Thus the random variables under consideration are determined *locally*.

We find two dimensional probabilities

$$\mathbf{P}(\omega \in \Omega : A^{(1)}(\omega) = \epsilon_1, B^{(1)}(\omega) = \epsilon_1') = \mathbf{P}(\epsilon_1, 0, \epsilon_1', 0) = \frac{1}{4}p_{11}(\epsilon_1, \epsilon_1'), \ldots,$$

$$\mathbf{P}(\omega \in \Omega : A^{(2)}(\omega) = \epsilon_2, B^{(2)}(\omega) = \epsilon_2') = \mathbf{P}(0, \epsilon_2, 0, \epsilon_2') = \frac{1}{4}p_{22}(\epsilon_2, \epsilon_2').$$

We also consider the random variables which are responsible for selections of pairs of ports to PBSs. For the device at LHS:

$$\eta_L(\epsilon_1, 0, 0, \epsilon_2') = \eta_L(\epsilon_1, 0, \epsilon_1', 0) = 1, \eta_L(0, \epsilon_2, 0, \epsilon_2') = \eta_L(0, \epsilon_2, \epsilon_1', 0) = 2.$$

For the device at RHS:

$$\eta_R(\epsilon_1, 0, \epsilon_1', 0) = \eta_R(0, \epsilon_2, \epsilon_1', 0) = 1, \eta_R(0, \epsilon_2, 0, \epsilon_2') = \eta_R(\epsilon_1, 0, 0, \epsilon_2') = 2.$$

---

[3]To match completely with Kolmogorov's terminology, we have to select a σ-algebra $\mathcal{F}$ of subsets of Ω representing events and define the probability measure on $\mathcal{F}$. However, in the case of a finite $\Omega = \{\omega_1, ..., \omega_k\}$ the system of events $\mathcal{F}$ is always chosen as consisting of all subsets of Ω. To define a probability measure on such $\mathcal{F}$, it is sufficient to define it for one-point sets, $(\omega_m) \to \mathbf{P}(\omega_m), \sum_m \mathbf{P}(\omega_m) = 1$, and to extend it by additivity: for any subset O of $\Omega, \mathbf{P}(O) = \sum_{\omega_m \in O} \mathbf{P}(\omega_m)$.

8.2.3 *Validity of CHSH-inequality for correlations taking into account randomness of selection of experimental settings*

Consider the correlations with respect to the probability $\mathbf{P}$ (which takes into account randomness of selections of experimental settings),

$$< A^{(i)}, B^{(j)} >= \int_\Omega A^{(i)}(\omega)B^{(j)}(\omega)d\mathbf{P}(\omega). \tag{8.9}$$

These are classical correlations for random variables taking values in $[-1, 1]$ and for them the CHSH-inequality $|S| \le 2$, see (8.2), holds (Theorem 1). Here S is defined by (8.1).

We remark that

$$< A^{(i)}, B^{(j)} >= \frac{1}{4} \Big[\sum_{\epsilon_1, \epsilon_1'=\pm 1} A^{(i)}(\epsilon_1, 0, \epsilon_1', 0))B^{(j)}(\epsilon_1, 0, \epsilon_1', 0))p_{11}(\epsilon_1, \epsilon_1')$$

$$+ \sum_{\epsilon_1, \epsilon_2'=\pm 1} A^{(i)}(\epsilon_1, 0, 0, \epsilon_2')B^{(j)}(\epsilon_1, 0, 0, \epsilon_2')p_{12}(\epsilon_1, \epsilon_2')$$

$$+ \sum_{\epsilon_2, \epsilon_1'=\pm 1} A^{(i)}(0, \epsilon_2, \epsilon_1', 0)B^{(j)}(0, \epsilon_2, \epsilon_1', 0)p_{21}(\epsilon_2, \epsilon_1')$$

$$+ \sum_{\epsilon_2, \epsilon_2'=\pm 1} A^{(i)}(0, \epsilon_2, 0, \epsilon_2')B^{(j)}(0, \epsilon_2, 0, \epsilon_2')p_{22}(\epsilon_2, \epsilon_2') \Big].$$

Let us fix some setting, e.g., $i = 1, j = 1$. Then taking into account the definitions of $A^{(1)}$ and $B^{(1)}$ we found that in the above expression only the first summand is nonzero.

Thus $< A^{(1)}, B^{(1)} >=$

$$\frac{1}{4} \sum_{\epsilon_1, \epsilon_1'=\pm 1} A^{(i)}(\epsilon_1, 0, \epsilon_1', 0))B^{(j)}(\epsilon_1, 0, \epsilon_1', 0))p_{11}(\epsilon_1, \epsilon_1') = \frac{1}{4} \sum_{\epsilon_1, \epsilon_1'=\pm 1} \epsilon_1\epsilon_1'p_{11}(\epsilon_1, \epsilon_1').$$

Thus classical correlations (taking into account randomness of setting selections) coincide (up to the factor $1/4$) with the correlations corresponding to the probability measures (p_{ij}):

$$C_{ij} \equiv \sum_{\epsilon_i, \epsilon_j'=\pm 1} \epsilon_i\epsilon_j'p_{ij}(\epsilon_i, \epsilon_j'), \tag{8.10}$$

namely,

$$C_{ij} = 4 < A^{(i)}, B^{(j)} > . \tag{8.11}$$

In particular, we can select the probabilities (p_{ij}) as the EPR-Bohm probabilities for the singlet state, see (8.3), then C_{ij} are the EPR-Bohm correlations used to violate the CHSH-inequality.

We stress that the *correlations C_{ij} are larger than classical ones*. In the general case,

$$C_{ij} = \frac{1}{\mathbf{P}(i,j)} < A^{(i)}, B^{(j)} > .$$ (8.12)

Thus randomization washes out a part of correlation. However, this washing effect is the probabilistic reality: in any CHSH-test experimentalists have to determine selection of experimental settings.[4]

8.2.4 *Quantum correlations as conditional classical correlations*

In classical probability theory one uses the notion of *conditional expectation* of a random variable (under the condition that some event occurred). This notion is based on Bayes' formula defining conditional probability, see Chapter 1. We shall use this notion to definite *conditional correlation* – under the condition that the fixed pair (i, j) of experimental settings is selected.

Let $(\Omega, \mathcal{F}, \mathbf{P})$ be an arbitrary probability space and let $\Omega_0 \subset \Omega$, $\Omega_0 \in \mathcal{F}$, $\mathbf{P}(\Omega_0) \neq 0$. We also consider an arbitrary random variable $\xi : \Omega \to \mathbf{R}$. Then the *conditional mathematical expectation* (average) of the random variable ξ, conditioned to the event Ω_0, is defined as follows:

$$E(\xi|\Omega_0) = \int_\Omega \xi(\omega) d\mathbf{P}_{\Omega_0}(\omega),$$

where the conditional probability $\mathbf{P}_{\Omega_0}$ is defined by Bayes' formula:

$$\mathbf{P}_{\Omega_0}(U) \equiv \mathbf{P}(U|\Omega_0) = \mathbf{P}(U \cap \Omega_0)/\mathbf{P}(\Omega_0).$$

In fact, this is the heart of our interpretation of violations of the Bell-type inequalities: one has to understand that from the viewpoint of classical probability theory these are inequalities based on conditional averages (correlations). Later we shall discuss this point in more detail.

---

[4]If one ignores this fact, then his description of the CHSH-test would not match the real experimental situation, because an important part of randomness involved in the test would be missed. In the typical quantum mechanical treatment of the Bell-type experiments randomness of context selection is not present explicitly. Only correlations for fixed contexts are considered. However, contribution of context selection can also be treated quantum mechanically, see section 8.4.

Let us come back to our unifying probability space. Take in the above definition

$$\Omega_0 \equiv \Omega_{ij} = \{\omega \in \Omega : \eta_L(\omega) = i, \eta_R(\omega) = j\}.$$

We remark that $\mathbf{P}(\Omega_{ij}) = 1/4$. The latter implies that

$$E(A^{(i)}B^{(j)}|\eta_L = i, \eta_R = j) = \int_\Omega A^{(i)}(\omega)B^{(j)}(\omega)d\mathbf{P}_{\Omega_{ij}}(\omega) \qquad (8.13)$$

$$= 4\int_{\Omega_{ij}} A^{(i)}(\omega)B^{(j)}(\omega)d\mathbf{P}(\omega) = 4\int_\Omega A^{(i)}(\omega)B^{(j)}(\omega)d\mathbf{P}(\omega) = 4 < A^{(i)}, B^{(j)} >$$

(since the product $A^{(i)}(\omega)B^{(j)}(\omega)$ is nonzero only on the set Ω_{ij}). Hence, we have:

$$E(A^{(i)}B^{(j)}|\eta_L = i, \eta_R = j) \equiv 4 < A^{(i)}, B^{(j)} > . \qquad (8.14)$$

By comparing (8.11) and (8.14) we obtain that the correlations C_{ij} which correspond to the collection of probabilities (p_{ij}) coincide with the classical conditional correlations, condition with respect to the choice of experimental setting.

Thus, we can identify the correlations C_{ij} obtained with the aid of probabilities (p_{ij}) with the corresponding conditional correlation for the unifying Kolmogorov space:

$$C_{ij} = E(A^{(i)}B^{(j)}|\eta_L = i, \eta_R = j). \qquad (8.15)$$

We were able to embed the correlations C_{ij} collected (separately) for in general incompatible experimental settings into the classical probability space. We remark that in principle by themselves these correlations can have the non-Kolmogorovian structure. It can happen that there is no a single classical probability space in which C_{ij} can be considered as unconditional correlations.

In particular, we can select the probabilities (p_{ij}) as the EPR-Bohm probabilities for the singlet state, see (8.3). Then we obtain the *representation of the corresponding quantum correlations as classical conditional correlations*.

Hence, the quantum correlations are present in our classical probability model for the CHSH-test, but they are not simply correlations: they are conditional correlations.

8.2.5 Violation of the CHSH-inequality for classical conditional correlations

As we have seen, the CHSH-inequality is satisfied for classical (unconditional) correlations (8.9).

Is there any reason to expect that it is also satisfied for classical conditional correlations?

The answer is no. In the classical probability model Theorem 1 can be proven only for unconditional correlations. Thus in principle conditional correlations can violate the CHSH-inequality. And this is not surprising from the Kolmogorovian viewpoint. Set

$$S_{\mathrm{C}} = E(A^{(1)}B^{(1)}|\eta_L = 1, \eta_R = 1) + E(A^{(1)}B^{(2)}|\eta_L = 1, \eta_R = 2) \quad (8.16)$$

$$+E(A^{(2)}B^{(1)}|\eta_L = 2, \eta_R = 1) - E(A^{(2)}B^{(2)}|\eta_L = 2, \eta_R = 2)$$

(here "C" is the abbreviation for conditional). By using the equality (8.14) we obtain that

$$S_{\mathrm{C}} = 4S. \quad (8.17)$$

Since by general Theorem 1 the quantity S is majorated by 2, the quantity S_{C} is majorated by 8. But the upper bound 8 is really too rough and, to obtain a better upper bound, we have to proceed more carefully.

Theorem 2. ("Strong CHSH-inequality") *Let $A^{(i)}, B^{(j)}, i, j = 1, 2$, be random variables defined as (8.5), (8.6), (8.7), (8.8). Then the corresponding combination of correlations S, see (8.1), satisfies the stronger version of the CHSH-inequality:*

$$|S| \leq 1. \quad (8.18)$$

Proof. Consider classical correlation $< A^{(i)}, B^{(j)} >= \int_{\Omega} A^{(i)}(\omega) \times B^{(j)}(\omega)d\mathbf{P}(\omega)$. We state again that the product $A^{(i)}(\omega)B^{(j)}(\omega)$ is nonzero only on the set Ω_{ij} and $\mathbf{P}(\Omega_{ij}) = 1/4$. We also use the condition that all random variables are valued in $[-1,1]$. Hence, in fact,

$$| < A^{(i)}, B^{(j)} > | = \left| \int_{\Omega_{ij}} A^{(i)}(\omega)B^{(j)}(\omega)d\mathbf{P}(\omega) \right| \leq \mathbf{P}(\Omega_{ij}) = 1/4.$$

Thus each correlation in the combination S of four correlations is bounded by 1/4. Hence, the inequality (8.18) holds.

We remark that, for this concrete Kolmogorov space, the CHSH-inequality is "trivialized"; in particular, the signs of correlations in S do not play any role. We remark that this theorem is valid for any collection of probabilities (p_{ij}), see (8.4).

Corollary 1. (CHSH-inequality for conditional correlations.) *Let* $A^{(i)}, B^{(j)}, i, j = 1, 2$, *be random variables defined as (8.5), (8.6), (8.7), (8.8). Then the corresponding combination of conditional correlation S_C satisfies the inequality:*

$$|S_C| \leq 4. \tag{8.19}$$

Take now probabilities given by (8.3), the EPR-Bohm probabilities for the singlet state. For them $S_C = 2\sqrt{2}$ and the inequality (8.19) is not violated.

8.3   Statistics: Data from Incompatible Contexts

Our previous analysis of the CHSH-experiment showed that one has to be very careful by operating with statistical data collected in incompatible "sub-experiments" of some "compound experiment." However, we do not claim that one cannot use such incompatible (from unconditional viewpoint) data to derive some statistical conclusions.

8.3.1   *Medical studies*

As was pointed by Richard Gill (email exchange) data collected for incompatible experimental contexts is widely used in statistics, especially in medicine:

"For instance we compare the health of non-smokers who are passively exposed to cigarette smoke (living with a smoking partner), with those who don't. The conclusion is that a lifetime of exposure roughly doubles your risk of lung cancer and other smoking related negative outcomes: doubled from something tiny, to something also tiny."

Two experimental contexts, C_1 : "living with a smoking partner", and C_2 : "not", are incompatible, we are not able to collect statistics by using the same person in both contexts. Nevertheless, we compare the probability distributions $p(\cdot|C_i), i = 1, 2$, to come to conclusions. Is such a practice acceptable from our viewpoint? Definitely yes. We can do everything with these contextual probabilities, but we should not forget that they are contextual, i.e., they are not so to say "absolute Kolmogorovian probabilities."

In the same way in the CHSH-experiment, we can compare the probabilities $p_{ij}(\epsilon, \epsilon')$ and say that, for one pair of angles (one experimental context) the probability of, e.g., the $(+, +)$-configuration is larger than for another pair (another context).

Even in medical experiments we can apply the construction of the underlying Kolmogorov space which was constructed for the CHSH-data. Only in this space we can work with probabilities by using the laws of classical probability theory, but absolute probabilities are less than contextual-conditional, this is the result of taking into account randomness of selections of the contexts C_1 and C_2. This should be remembered. And it seems that statisticians remembered this difference between "absolute" and conditional probabilities. (May be I was wrong, but then those who treated $p(\cdot | C_i), i = 1, 2$, as "absolute probabilities" might (but need not) get the problems which are similar to the problems generated by violation of Bell's inequality for quantum systems.)

8.3.2 *Cognition and psychology*

We also point out that, although in medical statistics the data from incompatible experimental contexts is widely used (as Richard Gill stressed), in another domain (where statistical methods also play the crucial role), namely, cognitive psychology and related studies in game theory and economics, the similar use of such data is considered as leading to ambiguous conclusions and paradoxes, see Chapter 9 for contextual probabilistic analysis of the situation.

This situation can be illustrated by contextual analysis of the games of the *Prisoner's Dilemma* type. From the contextual viewpoint, there are the following incompatible contexts: C_+ (C_-): one prisoner knows that another will cooperate (not) with her and C : he has no information about the plans of another prisoner [163]. In cognitive psychology and behavioral economics there was collected a plenty of statistical data for such games in different setups. As was pointed out, for psychologists comparison of such data led to paradoxes. We remark that non-Kolmogorovness of data collected for the contexts C_+, C_-, C is demonstrated via *violations of the formula of total probability*. This formula, as well as Bell's inequality, can be considered as a test of non-Kolmogorovness. I proposed to use this test in cognitive science, psychology, and economics in [7], [148]. The first experimental study was devoted to recognition of ambiguous figures, see Conte et al. [61].

Later professor of cognitive psychology J. Bussemeyer coupled violation of FTP with data collected in games of the Prisoner's Dilemma type and with violation of the *Savage Sure Thing Principle* [49], see also [50]. The latter axiomatizes the rationality of players, in particular, of agents acting at the market. Generally in cognitive psychology non-Kolmogorovness was coupled to *irrationality of players* or more generally the presence of some *bias*. We also remark that in cognitive psychology Bell's type inequalities are also used as statistical tests of non-Kolmogorovness, see Conte et al. [62], Asano et al. [15].

8.3.3 *Consistent histories*

The consistent histories interpretation of quantum mechanics provides a similar viewpoint on violation of Bell's inequality [104].

This approach (opposite to the majority of other approaches) is heavily based on the use of the Kolmogorov axiomatics of probability theory in quantum physics. The adherents of consistent histories proceed at the mathematical level of rigorousness, they "even" define explicitly the spaces of elementary events serving different quantum experiments [104] (which is very unusual for other approaches).

If a set of histories is consistent, then it is possible to introduce a probability measure on it, i.e., to use the Kolmogorov probability model. The crucial point is that in quantum physics there exist inconsistent families of histories, so quantum measurements cannot be described by a single probability space; quantum probability theory is non-Kolmogorovian. The violation of Bell's inequality is explained in the same way as in this chapter – as a consequence of the non-Kolmogorovian structure of experimental data collected for different experimental contexts. In particular, the nonlocality issue is irrelative (as well as in this chapter). However, the consistent histories approach stops at this point, i.e., recognition of non-Kolmogorovness. I proceeded further and showed that, in fact, quantum probabilities can be considered as the classical conditional probabilities and that the Kolmogorov model covers even the quantum probabilities, but conditionally.

Finally, we remark that in the consistent histories approach there are no counterfactuals as well as in our approach.

8.3.4 *Hidden variables*

Our model of conditional embedding of the quantum probabilities in the Kolmogorov model can be considered as an extension of the space of hidden variables to include random parameters generating the selections of experimental settings. Such a hidden variable depends on the parameters for the selections of angles at both labs. One can say that this hidden variable is nonlocal (although observed quantities are local). However, this nonlocal structure of a hidden variable reflects the nonlocal setup of the experiment, and nothing else. (We repeat once again that the experimental setup of the Bell-type tests is nonlocal from the viewpoint of classical physics. This classical nonlocality of the experimental setup has nothing to do with action at the distance.)

8.4 Contextuality of Bell's Test from the Viewpoint of Quantum Measurement Theory

We have shown that by taking into account the randomness of selection of experimental settings in the Bell-test it is possible to embed probabilistic data collected in experiments for the fixed settings into the classical probability space. This embedding is not straightforward. As was rightly shown by Bell, the direct identification of observables (used in various experimental contexts) with classical random variables is impossible. The observables are not unconditionally compatible. It is impossible to define their joint probability distribution. However, they are conditionally compatible, where conditioning is with respect to selection of the experimental settings.

In the light of the above considerations we do not claim that "Bell was wrong"; he was completely correct in the sense that there is no possibility to proceed classicality. However, Bell's classicality is related to the unconditional probabilistic compatibility. Thus he was right mathematically. But he was wrong from the physical viewpoint, because unconditional treatment of probabilities collected in the Bell test does not match the real physical situation which are intrinsically contextual: data is collected for a few incompatible experimental contexts. Now we shall show that the same conclusion one can obtain by proceeding not in the contextual probabilistic framework (Chapter 5), but even by applying solely *the standard quantum formalism of measurement theory*, due to von Neumann [250].

It will be shown that the quantum correlations of polarization observables used in Bell's argument have to be interpreted as *conditional* quantum correlations. The crucial point of coming considerations is that in the real experiments the complete set of measurements is not reduced to polarization-measurements. There are additional observables which contributions are not taken into account in the standard calculations of the Bell-correlations. If we take them into account and calculate the correlations properly, in the quantum framework, then we will see that the complete quantum correlations do not violate Bell's inequality – in contrast to the conditional quantum correlations, conditional upon the results of aforementioned additional quantum observables ("missed observables").

In fact, these observables are widely discussed in the literature, but in the philosophic terms. Now we want to formalize these terms by using the quantum measurement theory. Following tradition, we may speak about the problem of *freedom of choice* (sometimes also referred as experimenter's *free will*). Everybody, both Bellian and anti-Bellian, agrees that freedom of choice (free will) plays an important role in the Bell-type experiments. And this question is not only discussed, but it is seriously taken into account by experimenters, who play with various random generators in their experiments.

However, in the theoretical quantum mechanical formalization of these experiments there is nothing about the contribution of freedom of choice into quantum correlations. We want to take into account this contribution. We treat selections of experimental settings as measurements and perform the corresponding calculations. In this framework the standardly used polarization-correlations appear as *conditional correlations* with respect to experimenter's choices of experimental contexts. In short, we can say that experimenter's freedom of choice (free will) have to be represented in the rigorous mathematical framework by using quantum measurement theory.

We emphasize that the Bell inequality is proven (in the classical probability theory, Kolmogorov 1933) for unconditional probabilities and correlations, see Theorem 1 (section 1.3.2) and Theorem 1 (section 8.1). There is no reason to expect that it would be valid for conditional ones. From this viewpoint it is not surprising that it can be violated for conditional quantities. From our viewpoint, the main message of quantum violation of Bell's inequality is encoded in the *Tsirelson bound* $2\sqrt{2}$. The classical probability theory cannot explain why conditional probabilities of some physical model violate Bell's inequality precisely up to this probabilistic constant $2\sqrt{2}$. The classical theory gives for conditional probabilities the upper bound 4.

We remark that the Tsirelson bound is obtained by using the Hilbert space calculus of states and observables. One can wonder: What is behind this calculus? Of course, we know well that for some of readers this question is forbidden (QM is complete), for others the answer is known and it is negative (there is nothing behind QM - no hidden variables). However, some readers still believe that QM is incomplete (and that there is no even action at distance). For them, one of the most natural candidates for subquantum models is the classical field model ("Einstein's dream"), e.g., of the PCSFT-type (see Chapter 5). For those who believe in "subquantum fields" (in particular, the zero point field), the Tsirelson bound is a consequences of the wave feature of quantum systems. Therefore it would be interesting to derive it by using the classical field theory; it can be done, e.g., in the PCSFT-framework [167].

8.5 Inter-relation of Observations on a Compound System and its Subsystems

8.5.1 *Averages*

Consider a compound system $S = (S, S')$, where S and S' have the state spaces H and K, respectively; thus S is represented in the state space $\mathcal{H} = H \otimes K$. Let $A_j, j = 1, ..., k$, be a group of observables on S, they are represented by Hermitian operators acting in H which are denoted by the same symbols. In general these observables are incompatible. Consider also an observable G on S', having the values $j = 1, ..., k$; it is represented by a Hermitian operator in K, $G = \sum_j jP_j$, where (P_j) is its spectral family consisting of mutually orthogonal projectors.

For any state ρ, a density operator in H, we can define the averages of observables $A_j : M_j = \mathrm{Tr}\rho A_j$, and, for any state σ, a density operator in K, the averages of observables $P_j : g_j = \mathrm{Tr}\sigma P_j$.

Now we consider the observables $\mathcal{A}_j$ on the compound system S

$$\mathcal{A}_j = A_j \otimes P_j. \tag{8.20}$$

Consider a state R of S. We can define the averages of the observables $\mathcal{A}_j$ for this state:

$$m_j = \mathrm{Tr}R\mathcal{A}_j. \tag{8.21}$$

Let the state of the compound system S be factorizable, i.e., its subsystems are not entangled,

$$R = \rho \otimes \sigma. \tag{8.22}$$

Then

$$m_j = M_j g_j. \tag{8.23}$$

Suppose now that by experimenting with the compound system S one "forgot" about the presence of the subsystem S'.

This forgetfulness has an interesting probabilistic effect: it induces the increase of averages, from m_j to M_j with the scaling coefficient $k_k = 1/g_j$.

Thus if one treats an experiment on the compound system S as an experiment on its proper subsystem S, the averages and probabilities can increase essentially. For example, let $\sigma = I/\dim K$. Then $g_j = \dim P_j/\dim K$ and $k_j = \dim K/\dim P_j$. We are, in fact, interested in the case $\dim K = 4$ and $\dim P_j = 1$, i.e., $k_j = 4$, the *four times increase of the magnitudes* of the averages and probabilities.

This increase of averages explains "the mystery of violations of Bell's type inequalities and superstrong quantum correlations".

8.5.2 *Correlations*

Now move to the case of quantum correlations. Let now $H = H_1 \otimes H_2$, i.e., S is by itself a compound system $S = (S_1, S_2)$, and let $K = K_1 \otimes K_2$, i.e., S' is by itself a compound system $S' = (S'_1, S'_2)$. For the state space H_1, we consider a pair of observables A_0, A_1 and, for the state space H_2, a pair of observables B_0, B_1; for K_1, a pair of observables represented by orthogonal projectors P_0, P_1 and, for K_2, a pair Q_0, Q_1. Finally, let ρ and σ be the states represented by density operators acting in $H = H_1 \otimes H_2$ and $K = K_1 \otimes K_2$.

In Bell's experimental scheme the observables in H_i represent polarization measurements and the observables in K_i represent measurements of outputs of random generators. The state ρ is the state of a pair of entangled photons $S = (S_1, S_2)$ and the state σ is a separable state of the pair of random generators, where the state of each random generator can (but need not) be given by a classical statistical mixture of two possible outputs. Of course, our scheme works for observables and random generators with an arbitrary number of outputs. By restricting the numbers of outputs to two we just try to keep closer to our concrete aim – Bell's experimental scheme.

We introduce correlations in H

$$C_{ij} = \mathrm{Tr} A_i \otimes B_j \rho = \mathrm{Tr} O_{ij}\, \rho, \tag{8.24}$$

where $O_{ij} = A_i \otimes B_j$, and in K

$$g_{km} = \mathrm{Tr} P_k \otimes Q_m \sigma = \mathrm{Tr} O'_{km}\, \sigma, \tag{8.25}$$

where $O'_{km} = P_k \otimes Q_m$.

Now we consider the state space of the compound system $S = (S, S')$ given by $\mathcal{H} = H \otimes K = H_1 \otimes H_2 \otimes K_1 \otimes K_2$. For any state given by a density operator R on $\mathcal{H}$, we can find the correlation of the observables given by the operators O_{ij} and O'_{km} :

$$c_{ij,km} = \mathrm{Tr} O_{ij} \otimes O'_{km} R = \mathrm{Tr} A_i \otimes B_j \otimes P_k \otimes Q_m R. \qquad (8.26)$$

Suppose that states in H and K are not entangled, i.e., $R = \rho \otimes \sigma$. Then

$$c_{ij,km} = \mathrm{Tr} O_{ij} \rho \, \mathrm{Tr} O'_{km} \sigma = C_{ij} \, g_{km}. \qquad (8.27)$$

Suppose again that by experimenting with the compound system S one "forgot" about the presence of the subsystem S'. If the correlation $g_{km;\sigma} < 1$, then:

This forgetfulness induces the increase of correlations, from $c_{ij,km}$ to $C_{ij;\rho}$.

In this way one obtain "superstrong nonclassical Bell correlations".

8.5.3 *Towards proper quantum formalization of Bell's experiment*

Consider the case of the two dimensional spaces K_1 and K_2. The corresponding bases in $K_t, t = 1, 2$, are $(|0\rangle_t, |1\rangle_t)$. To shorten the notation, further we omit the index t. Let $P_\alpha, Q_\alpha = |\alpha\rangle\langle\alpha|, \alpha = 0, 1$. Let the states in K_1 and K_2 be not entangled, i.e., $\sigma = \sigma_1 \otimes \sigma_2$ and each state σ_i is the "classical mixture":

$$\sigma_1 = p_0|0\rangle\langle 0| + p_1|1\rangle\langle 1|, \ \ \sigma_2 = q_0|0\rangle\langle 0| + q_1|1\rangle\langle 1|, \qquad (8.28)$$

where $p_0 + p_1 = 1$ and $q_0 + q_1 = 1$ and nonnegative. Then

$$g_{km} = p_k q_m < 1.$$

In particular, if all probabilities are equal, we obtain that

$$g_{km} = 1/4. \qquad (8.29)$$

This imply 4-times increase of correlations as the result of "missing" the subsystem S'.

In the Bell-type experiments, e.g., for the CHSH-inequality, one operates with the linear combination of correlations

$$C = C_{00} + C_{01} + C_{10} - C_{11}. \qquad (8.30)$$

It is convenient to represent C as the average of a single observable represented as the operator

$$\Gamma = O_{00} + O_{01} + O_{10} - O_{11} = A_0 \otimes B_0 + A_1 \otimes B_0 + A_0 \otimes B_1 - A_1 \otimes B_1. \quad (8.31)$$

Thus

$$C = \mathrm{Tr}\rho\Gamma. \quad (8.32)$$

However, one can proceed with this operator only by ignoring the second subsystem of the compound system $S = (S, S')$. By considering measurement on S and by taking into account correlations of observables on S' we come to the representation of the corresponding modification of the correlation function C in the following operator form:

$$\gamma = O_{00} \otimes O'_{00} + O_{01} \otimes O'_{01} + O_{10} \otimes O'_{10} - O_{11} \otimes O'_{11}. \quad (8.33)$$

The corresponding correlation function is given by the average

$$c = \mathrm{Tr}\rho \otimes \sigma \, \gamma = c_{00} + c_{01} + c_{10} - c_{11}, \quad (8.34)$$

where

$$c_{ij} = C_{ij}g_{ij}. \quad (8.35)$$

In the case of equal probabilities p_i, q_j, see (8.29), we have:

$$c = C/4. \quad (8.36)$$

Thus by taking into account that the second system S' is also involved in the experiment, we find that the "Bell correlation" function C decreases four times.

In particular, for the EPR-Bell correlations the Tsirelson bound for C, namely, $C_{\max} = 2\sqrt{2}$, which is the bound for conditional quantum correlations, leads to the following bound for unconditional quantum correlations

$$c = \frac{\sqrt{2}}{2} < 2. \quad (8.37)$$

8.6   Quantum Conditional Correlations

Consider the same compound system $S = (S, S')$ as in section 8.5.2. We now remark that the joint measurement of the observables mathematically represented by the operators $O_{ij} \otimes I$ and $I \otimes O'_{km}$ can be treated as their sequential measurement. The correlations are the same. Now let us consider another problem – to find (again with the aid of the quantum theory of measurement) *conditional correlations* of the observables A_i and B_j -

conditioned to the fixed outputs of the measurements of the observables given by P_i and Q_j.

First we measure (on the subsystem S' of S) the observable $O'_{ij} = P_i \otimes Q_j$. In the quantum formalism this can be treated as measurement on S of the observable given by $I \otimes O'_{ij}$. If the initial state of S is R, then by getting the result $(P_i = 1, Q_j = 1)$ we know that the initial state is transformed to the post-measurement state:

$$R \to R_{ij} = \frac{(I \otimes O'_{ij})R(I \otimes O'_{ij})}{\text{Tr}(I \otimes O'_{ij})R(I \otimes O'_{ij})}. \qquad (8.38)$$

Now in accordance with the quantum formalism for conditional measurements we measure the observable $O_{ij} \otimes I$. The corresponding (conditional) average is given by

$$C_{ij|\text{cond}} = \text{Tr}R_{ij}(O_{ij} \otimes I). \qquad (8.39)$$

If $R = \rho \otimes \sigma$, then $(I \otimes O'_{ij})R(I \otimes O'_{ij}) = \rho \otimes O'_{ij}\sigma O'_{ij}$. In particular, the denominator in (8.38) is given by $\text{Tr}O'_{ij}\sigma O'_{ij}$. We also have:

$$C_{ij|\text{cond}} = \frac{\text{Tr}\rho O_{ij} \, \text{Tr}O'_{ij}\sigma O'_{ij}}{\text{Tr}O'_{ij}\sigma O'_{ij}} = \text{Tr}\rho O_{ij}. \qquad (8.40)$$

Hence, the term $\text{Tr}O'_{ij}\sigma O'_{ij}$ disturbing Bell's argument peacefully disappeared.

Thus, if one treats the correlations C_{ij} in Bell's correlation function C, see (8.31), as the quantum *conditional correlations*, then the diminishing effect discussed in section 8.5.3 disappears.

Thus by using the standard quantum formalism we demonstrated that complete quantum correlations in Bell's type experiment do not violate Bell's inequality. It seems that Bell and following him scientists used improper quantum mechanical description of such experiments. The conditional quantum correlations, where conditioning is to the choice of fixed experimental settings (e.g., the orientations of PBSs), were compared with unconditional classical correlations.

From our analysis, it is clear that in Bell's framework there are two justified ways of proceeding:

- either with conditional quantum correlations (which are used in the literature on Bell's argument) and then compare them with classical conditional correlations,
- or with complete (unconditional) quantum correlations and then compare them with classical unconditional correlations.

In the first case, Bell's argument is collapsed, since classical conditional correlations can violate Bell's inequality; in the second case, it is collapsed, since complete (unconditional) quantum correlations satisfy Bell's inequality.

In contrast to Feynman [88], Bell missed the contextual probabilistic structure of the test on violation of Bell's inequality. This led to unconditional treatment of the EPR-Bohm correlations in such tests with all aforementioned consequences of such treatment.

8.7   Classical Probabilistic Realization of "Random Numbers Certified by Bell's Theorem"

The idea that quantum randomness differs crucially from classical randomness is due to von Neumann who pointed out that the first one is intrinsic and the second one is reducible to variability in an ensemble (Chapter 6). At the first stages of the development of QM von Neumann's viewpoint on quantum randomness and its exceptional nature was merely of the purely foundational value. It was used to support various no-go statements, starting with von Neumann's no-go theorem [250]. However, with the development of quantum information theory and its technological implementation the interrelation between classical and quantum randomness started to play the important role in real applications.

In this section we shall discuss one very novel and important application of quantum randomness, namely, in the form of random number generators certified by Bell's theory, see, e.g., [217] for details and review. As was claimed in [217]:

"It is thereby possible to design a new type of cryptographically secure random number generator which does not require any assumption on the internal working of the devices. This strong form of randomness generation is impossible classically and possible in quantum systems only if certified by a Bell inequality violation."

This statement is in contradiction with the classical probabilistic model for quantum probabilities and the corresponding correlations violating the CHSH inequality which was constructed in the previous sections. By using this classical probabilistic representation of CHSH-probabilities, we can now construct the classical random generator (at least theoretically) producing 6-dimensional vectors

$$\xi_m = (A_m^{(1)}, A_m^{(2)}, B_m^{(1)}, B_m^{(2)}, \eta_L, \eta_R), m = 1, 2, \ldots.$$

The last two coordinates take values 1,2 (selection of experimental settings) and the first four coordinates take the values $\{-1, 0, +1\}$. Now if we filter from these 6-dimensional vectors the zero coordinates, we shall get 4-dimensional vectors

$$(a_m, b_m, \eta_{Lm}, \eta_{Rm}) \tag{8.41}$$

which are indistinguishable from the viewpoint of Bell's test from those obtained, e.g., in the quantum optics experiments. The corresponding conditional probabilities, conditioning with respect to selection the fixed pairs of settings (θ_i, θ'_j), violate the CHSH inequality.

Thus we can apply to the string of data (8.41) the standard approaches to randomness based on classical probability space (Chapter 2).

Finally, we remark that we do not question the quality of "Random Numbers Certified by Bell's Theorem". If they pass the standard tests for randomness, e.g., the NIST-test, they are good, if not, then they are bad, regardless of violation of the CHSH-inequality and regardless of the problem of closing of experimental loopholes. The main concern is about the claims that quantum random generators are in some way better than classical ones.

Chapter 9

Quantum Probability Outside of Physics: from Molecular Biology to Cognition

Recent years were characterized by the tremendous extension of the domain of applications of the quantum probability (QP). The operator representation of observables and states is fruitful not only in quantum physics but also for applications outside of physics, the keywords are *non-Kolmogorovness and simplicity.*

On one hand, in practically any domain of science we can find data which cannot be embedded in common classical probability space (at least explicitly, i.e., without conditioning and post-selection, cf. Chapters 5 and 8). This is a consequence of contextuality of data, i.e., dependence on contexts of measurements. And in general there are no reasons to expect compatibility of contexts and, hence, of data. It is *convenient* to describe such data with the aid of QP (or other operational probabilistic models).

On the other hand, in many situations the measure-theoretic approach to probability is technically more complicated (and sometimes essentially) than QP. As we have seen, e.g., in Chapter 1 and especially in Chapter 3, the rigorous presentation of the measure-theoretic reasoning can be complex and moreover based on some assumptions, such as, e.g., the axiom of choice, which mathematical justification is really fuzzy. In fact, the axiomatic set theory is not less "mysterious" than QM. And the mathematical apparatus of QM, linear algebra, is definitely simpler than the measure theory (and the standard scenarios of quantum information theory is written for finite-dimensional spaces)[1].

---

[1]The construction of the Kolmogorov probability space for data collected in Bell's test, section 8.2.2, can be used for any collection of incompatible contexts in general. The possibility to construct a classical probability space demystifies contextuality and, in particular, QM. However, this does not diminish the role of QP as an operational probabilistic formalism. The quantum formalism provides the integral representation covering all possible "quantum contexts". Our construction of Kolmogorovization of

In this chapter we shall discuss applications of QP in biology. Here biology is treated widely as even covering cognition and its derivatives: psychology and decision making, sociology and behavioral economics and finances.

Section 9.1 is devoted to a general discussion about the use of QP in biological modeling; in section 9.2 we oppose our *quantum-like* (QL) modeling of bio-phenomena to conventional quantum biophysics; in section 9.3 our studies are coupled to the information interpretation of QM and especially QBism; in section 9.3.3 we briefly present author's mathematical model of information reality [127] which is based on departure from "physical space" (mathematically modelled as the real continuum) to "information space" (mathematically modeled as a hierarchic tree; e.g., p-adic space [127], [246], [126]). This first block of sections has the foundational character and in principle the reader can jump directly to section 9.4 motivating the use of generalized quantum observables, *open quantum systems* and more generally *quantum adaptive dynamics* in biology.

9.1   Quantum Information Biology

We discuss a novel field of research, *quantum information biology* (QIB). It is based on application of a recently developed quantum information formalism [9] - [16], [28] *quantum adaptive dynamics*, outside of physics.[2] Quantum adaptive dynamics describes in the most general setting mutual adaptivity of information states of systems of any origin (physical, biological, social, political) as well as mutual adaptivity their co-observations, including self-observations (such as performed by the brain).

Nowadays *quantum information theory* is widely applied to quantum computers, simulators, cryptography, and teleportation. Quantum infor-

---

contextual probabilities has to be concrete for any particular case. And from this point of view Bell's test is not so bad: there are only four contexts which have to be unified. However, in many applications the number of incompatible contexts involved can be huge, see, for example, [5], where the Kolmogorov model for the experiment on recognition of an ambiguous figure is very complex; here we have to unify $n!$ contexts and n is sufficiently large. In such situations the existence of the classical probability model has merely foundational value. The use of QP simplifies models essentially.

[2]QIB is not about quantum physics of bio-systems (see [8] for extended review), in particular, not about quantum physical modeling of cognition, see section 9.2 for details. Another terminological possibility would be "quantum bio-information". However, the latter term, bio-information, has been reserved for a special part of biological information theory studying computer modeling of sequencing of DNA. We do not want to be associated with this activity having totally different aims and mathematical tools.

mation theory is fundamentally based on quantum probability (QP).

We noticed that biological phenomena often violate the basic laws of classical probability (CP) [151], [154]; in particular, the formula of total probability (FTP). Therefore in biology it is natural to explore one of the most advanced and well known nonclassical probability theories, namely, QP. This leads to the quantum information representation of biological information flows which differs crucially from the classical information representation.

Since *any bio-system S is fundamentally open* (from a cell to a brain or a social system), i.e., it cannot survive without contacts with environment (physical, biological, social), it is natural to apply the powerful and well developed apparatus of theory of open quantum systems [210] to the description of biological information flows. Adaptivity of the information state of S to an environment is a special form of bio-adaptivity: mutual adaptivity of information states of any pair of bio-systems S_1 and S_2. Since quantum information theory provides a very general description of information processing, it is natural to apply it to model adaptive dynamics. This led us [11] to quantum adaptive dynamics. Surprisingly we found that such an *operational quantum formalism* can be applied to any biological scale by representing statistical experimental data by means of quantum probability and information. During the last years a general model representing all basic information flows in biology (from molecular biology to cognitive science and psychology and to evolution) on the basis of quantum information/adaptive dynamics was elaborated. In a series of our works [9] - [16], [28] the general scheme of embedding of biological information processing into the quantum information formalism was presented and the foundational issues related to the usage of quantum representation for macroscopic bio-systems, such as genome, protein,...., brain,...., bio-population were discussed in details and clarified.

Our theory can be considered as the informational basis of *decision making*. Each bio-system (from a cell to a brain and to a social or ecological system) is permanently in process of decision making. Each decision making can be treated as a self-measurement or more generally as an adaptive reply to signals from external and internal environments.

On this basis, we believe that QIB is the most predictive tool to know our future state on earth. We expect that *this quantum-like operator formalism is a kind of brave trial to unify our social and natural sciences.*

9.2   Inter-relation of Quantum Bio-physics and Information Biology

This section is devoted to comparison of quantum bio-physics with QIB. Those who have been satisfied with a brief explanation given in section 9.1 can skip this section.

First of all, we emphasize that our applications of the methods of QM to modeling of information flows in bio-systems have no direct relation to *physical quantum processes* in them, cf. [8]. For example, by considering quantum information processing by a cell or by a protein molecule [9] - [16], [28] we do not treat them as quantum *physical systems.* In fact, already a single cell or even a protein molecule is too big to be treated in the conventional QM-framework which was elaborated for microscopic systems. (N. Bohr emphasized the role of the fundamental quantum of action - the Planck constant h.) Of course, in the quantum foundational community there have been a lot of discussion on the so-called macroscopic quantum phenomena. However, it seems that the magnitudes of some important physical parameters of bio-systems do not match with scales of macro-quantum phenomena (e.g., Tegmark's [241] critique of "quantum mind", cf. with [8]). For example, bio-systems are too hot comparing with Bose-Einstein condensate or Josephson-junction. At the same time we can point to experiments of A. Zeilinger continued by Arndt, see [8], on interference for macro-molecules, including viruses. They were performed at high temperature. We can also point to recent studies claiming the quantum physical nature of the photo-synthesis.

In any event in QIB we are not interested in all complicated problems of macro- and, in particular, bio-quantumness. In particular, by applying our formalism to cognition we escape involvement in hot debates about a possibility to treat the hot and macroscopic brain as processing quantum physical (entangled) states, since we do not couple our model to the "quantum physical brain project", e.g., theories of Penrose [216] and Hameroff [105]. For example, the complex problem of whether micro-tubules in the brain are able to preserve quantum physical entanglement (before its decoherence) for a sufficiently long time, e.g., to perform a step of quantum computation, is outside of our theory. QIB describes well processing of mental informational states, independently of who will be right: Penrose and Hameroff or Tegmark. (It is also possible that this debate will continue forever.) The ability of a bio-system to operate with superpositions of information states including brain's ability to form superpositions of mental states is the key-issue.

For a moment, there is no commonly accepted model of creation of superposition of information states by a bio-system, in particular, there is no proper neurophysiological model of creation of mental superposition. For the latter, we can mention a series of attempts to model states superposition with the aid of classical oscillatory processes in the brain, as it was proposed by J. Acacio de Barros and P. Suppes [1], including classical electromagnetic waves in the brain, see, e.g., for a model [164] proposed by A. Khrennikov. In these studies mental states are associated with classical physical waves or oscillators, the waves discretized with the aid of thresholds induce the probabilistic interference exhibited in the form of violations of FTP, cf. Chapter 5.

In general we consider violations of FTP by statistical data collected in bio-science, from molecular biology to cognitive psychology, as confirmation of the ability of bio-system to operate with state superpositions, cf. with QM, section 5.1. In an experiment, in QM as well as in biology, we do not detect superposition of "individual information states" such as superposition of two classical waves. *Informational states of bio-systems are probabilistic amplitudes.* In QIB we discuss the ability of bio-systems to represent probabilities by vector amplitudes and operate with such amplitudes by using the operations which are mathematically represented as the matrix calculus. From this operational viewpoint an electron does not differ so much from a cell, or a brain, or a social system.

In section 5.4, see also [163], it was shown that, for probabilistic data of any origin, violation of FTP leads to representation of states by probability amplitudes - *the constructive wave function approach.* It might be that bio-systems really use this algorithm of production of amplitudes from probabilities – *the quantum-like representation algorithm.* In a bio-system probability can be treated as frequency. And frequencies can (but need not) be encoded in oscillatory processes in bio-systems.

Unfortunately, the existing quantum-like algorithm works only for dichotomous observables. But there is still a hope that in future a similar algorithm will be designed even for non-dichotomous observables. On the other hand, may be the brain and other bio-processors of information operate solely with dichotomous "yes"-"no" observables creating from them complex information patterns. In such a case the existing algorithm of representation of data by amplitudes or in the abstract framework by vectors encoding probabilities can be considered as a candidate for the linear space encryption of information in biological information processors. This *linear structure of QL-representation of probabilities* is one of the main distinguishing features of the QL model of cognition. Linearization is always simplification and one cannot exclude that bio-systems developed evolutionary the ability to represent probabilities by vectors and process probabilistic data, e.g., in the process of decision making, by using linear dynamics. We can speculate that the brain could, in principle, use a part of its *spatial representation* "hardware" for the vector representation of probabilities and, hence, for decision making and processing of information at the advanced level. For a moment, this is pure speculation.

9.3   From Information Physics to Information Biology

9.3.1   *Operational approach*

In QIB, similarly to quantum physics we treat the quantum formalism as an *operational formalism* describing (self-)measurements performed by bio-systems.

Neither QM nor biology (in particular, cognitive science) can explain why systems under study produce such random outputs - violating the CP-laws. Moreover, in QM, according to the Copenhagen interpretation, it is in principle impossible to provide some "explanation" of quantum random behavior, e.g., by using a more detailed description with the aid of *hidden variables*. In spite of this explanatory gap, QM is one of the most successful scientific theories. One may hope that a similar operational approach would finally lead to creation of a novel and fruitful theory for bio-systems.

In principle, in QIB it is also possible to proceed without the assumption about irreducible randomness and, hence, without looking for its biological sources. On the other hand, it is important to pose the following stimulating question:

Is an individual bio-system intrinsically random?

9.3.2   *Free will problem*

In principle, for human beings, applications of QIB to cognition and decision making, one can couple the intrinsic quantum(-like) randomness with *free will*. In the recent debates on non/deterministic nature of quantum processes, the role of non/existence of free will was actively discussed. At the moment two opposite positions (no/yes) aggressively coexist.

We emphasize that the rejection of the existence of free will, the position of G. 't Hooft [242] known as *super-determinism*, opens a possibility to treat QM as a deterministic theory, e.g. [243], [244]. We also remark that a part of quantum foundation community thinks that the issue of non/existence of free will is irrelevant to the problem of quantum (in)determinism.

Exploring the free will assumption may be helpful to motivate the presence of *intrinsic mental randomness* and in this way to couple QIB even closer to Zeilinger-Brukner (and von Neumann) interpretation. However, by proceeding in this way one has to be ready to meet a new problem of the following type:

Have a cell, DNA or protein molecular a kind of free will?

We can formulate the problem in another way around:

Is human free will simply a special exhibition of intrinsic randomness which is present at all scales nature?

As we have seen in previous chapters the problem of interrelation of classical and quantum probabilities (as well as randomnesses) is extremely complex. In some way it is easier for probabilities, since here we have satisfactory mathematical models, e.g., the Kolmogorov model [177] of classical probability. However, as we have seen (Chapter 3), even in the classical case a commonly acceptable notion of randomness has not yet been elaborated, in spite of tremendous efforts during the last century. In the quantum case (Chapter 7) the situation is even worse.

9.3.3 Bohmian mechanics on information spaces and mental phenomena

We point out that the purely information approach to mental phenomena and their unification with natural phenomena was developed in a series of works, see, e.g., [128] - [130], [7, 131, 138, 149], [148].

In particular, a version of Bohmian mechanics on information spaces [129], [148] was explored to model mental processes, especially unconscious processing of information and transition from unconscious to conscious. Some Bohmians, e.g., D. Bohm by himself and B. Hiley, interpreted the wave function as *a field of information* [36]. However, for them this information field was still defined on the physical space, because, for D. Bohm, as well as his predecessor L. De Broglie, the physical space was the basis of reality. In my works mental information fields were defined on "information space" reflecting the hierarchic tree-like representation of information by cognitive systems. Mathematically such mental trees can be represented with the aid of so called p-adic numbers [162], [81]. The brain can generate such information space as the result of functioning of hierarchically organized neuronal trees [149, 150].

I advertised the ideology that the information space plays the primary role both in mental and natural phenomena. The laws of information dynamics are the most fundamental laws of nature. Laws describing behavior of material structures (including physical fields) are so to say shadows of the information laws (cf. with Plato's world of ideas).

It seems that in real brain's functioning we observe the opposite picture: neuronal trees (material structures) generate "mental fields". However, in the information reality framework the processes mapped to the information space induce the corresponding material structures. In particular, neuronal trees growing in our brains are "shadows" of the corresponding information structures – "mental trees" ecoding information hierarchically [149, 150]. Of course, although in my approach information was treated as the most fundamental entity of nature, so similarly to the views of Zeilinger-Brukner, Fuchs-Schack-Caves-Mermin, D'Ariano-Chiribella-Perinotti, I am not sure that these authors would support my purely information model coupling the mental and physical phenomena.

This (pure information reality) ideology is sufficiently fruitful in modeling of mental phenomena: unconscious↦conscious transition, psychoanalysis, depression, sexuality [7, 131, 138, 149], [148]. However, later I became less addicted to the idea of information reality.

The main problem of the information interpretation of QM and more generally purely information representation of nature is to assign a proper meaning to the notion of information.

It is very difficult to do; one has to define information not about some real objectively existing stuffs, but information as it is by itself. Of course, one can simply keeps to Zeilinger's position (private discussions): information is a primary notion which cannot be expressed through other "primitive notions". However, not everybody would be satisfied by such a solution of this interpretational problem.

Although, as was pointed out, may be founders of the information interpretation of QM would not support my model of purely information reality, they also should solve the problem of meaning assignment for the notion of information, cf. with Mermin's work "Whose knowledge?" [201].

My universe is populated by *information processors* [7, 131, 138, 149], [148], humans, cells, stones, electrons. Whose information? The answer is "of these information processors". Information – about what? The answer is "about information". Well, it is not clear whether the invention of information processors was really a step forward comparing with Zeilinger's "information as the primary notion."

9.3.4 *Information interpretation is biology friendly*

In any event the information interpretation matches perfectly with our aims, although its creators may not support the attempts to apply it outside of physics. (We remark that the position of A. Zeilinger is quite supporting for those who try to explore the operational approach to cognition.[3]) There is a chance that physicists using this interpretation would welcome the born of quantum cognition and more generally QIB. There was no any chance to find understanding of physicists keeping to the orthodox Copenhagen interpretation – the wave function as the most complete representation of the *physical state* of an individual quantum system such as an electron.

9.4 Nonclassical Probability? Yes! But, Why Quantum?

One may say: "Yes, I understood that the operational description of information processing in complex bio-systems can be profitable (independently from a possibility to construct a finer, so to say hidden variables, theory). Yes, I understand that bio-randomness is nonclassical and it is not covered by CP and, hence classical information theory cannot serve as the base for information biology. But, why do you sell one concrete nonclassical probability theory, namely, QP and the corresponding information theory? May be (as Zeilinger guessed) QP and quantum information serve well for a special class of physical systems and their application to biology may meet hidden pitfalls?" Yes, we agree that there is a logical gap: nonclassical does not imply quantum. Negation of CP is not QP. (Of course, here we proceed quite primitively by ignoring the complex inter-relation between classical and quantum probability and randomness. To be more explicit, we have to have in mind that QP can be treated as a special calculus of conditional classical probabilities, Chapter

---

[3] In a series of private discussions with the author of this book, he expressed satisfaction that by exploring the ideology of QM for modeling of cognition one finally breaks up the realist attitude dominating in cognitive science and modeling of brain's functioning and based on attempts to reduce mental phenomena to firing of neurons. At the same time he was not sure at all that, to model cognition, one has to use *precisely the quantum formalism*: novel operational formalisms representing information processing might be more appropriate, see also section 9.4. A. Zeilinger has amazingly powerful intuition and his remark has to be taken seriously.

5. As was already pointed out, this calculus is very convenient, in particular, because of its linearity. Therefore, although one can always proceed with conditionally compatible observables (Chapters 5 and 8), the use of the latter might induce additional complications (including mathematical), because it destroys the linear structure of the probabilistic representation).

Once again, we ask: Why exploring QP is so attractive? First of all, because it is the most elaborated nonclassical probability theory. It is attractive to use its advanced methods. One may expect that some basic nonclassicalities of random responses of bio-systems can be described by the standard quantum formalism. This strategy was very fruitful, especially at the initial stage of development [130], [145], [133], [148], [61], [62], [163], [50], [49], [106], [1], [22], [90], [83], [84], [9] - [16], [172], [173], [166], [51], [5].

Now the crucial question arises:

Can the standard quantum formalism cover completely information processing by bio-systems?

Here "standard" refers to Schrödinger's equation for the state evolution, the representation of observables by Hermitian operators and the von Neumann-Lüders projection postulate for quantum measurement.

Our answer is "no". Already in quantum information one uses the open system dynamics and generalized observables given by POVMs. It is natural to expect that they can also arise in bio-modeling. As was found by the author [163], all matrices of transition probabilities appeared in QIB are not *doubly stochastic*. However, the representation of observables by Hermitian operators with non-degenerate spectra leads to double stochastic matrices, section 4.3. Thus one cannot proceed with von Neumann-Lüders observables, at least with nondegenerate spectra.

In applications outside QM, it is very convenient to work with observables having nondegenerate spectra. For such observables, the range of values determines the dimension of the state space. Otherwise the problem of determination of the state space dimension would be very difficult. Here we do not have, e.g., the group-theoretical methods to determine the dimensions of representations. In bio-applications, in particular, for cognition, by assuming that spectrum can be degenerate one really opens Pandora's box. QIB is a phenomenological theory. Therefore it is difficult, if possible at all, to assign some meaning to additional dimensions appearing as the result of splitting of the integral dimensions corresponding to the observed spectrum. In such a situation it is more natural to proceed with POVMs in state spaces of the dimensions corresponding to the ranges of values of observables.

Another argument for departure from the standard quantum formalism is that in the *constructive wave function approach,* i.e., construction of complex probability amplitudes from data of any origin violating FTP [163], in general observables are represented not by Hermitian operators, but by POVMs and even by generalized POVMs which do not sum up to the unit operator [163] (this problem was ignored in Chapter 5, where considerations were restricted to the case of doubly stochastic matrices of transition probabilities).

A strong argument against the use of solely Hermitian observables in cognitive science was presented in [166]: it seems that the Hermitian description of the order effect for a pair of observables A, B, i.e., disagreement between $A - B$ and $B - A$ probabilities, is incompatible with $A - B - A$ respectability: first A-measurement, then B-measurement, then again A-measurement and the first and the last values of A should coincide with probability 1. At the same time the standard opinion polls demonstrating the order effect have the property of $A - B - A$ respectability. However, it seems that this problem is even more complicated: even the use of POVMs cannot help so much [166]. It seems that we have to go really beyond the quantum measurement formalism, beyond theory of quantum instruments. One of such novel generalizations was proposed in the series of works [11], [16] - quantum adaptive dynamics.

Appendix A

Växjö Interpretation-2002

Here we present the pioneer paper [143] on the Växjö interpretation of QM. It had the strong anti-Copenhagen attitude. It emphasized the possibility of the *realist interpretation of QM* by criticizing Bohr's argument against such an interpretation based on the principle of complementarity. The main question posed in this paper can be formulated as:

Why should complementarity contradict to the realist interpretation?

We omit the introduction to the paper [143] and start with its section 2; we comment some statements made in 2002 in footnotes by marking such footnotes as "Comment."

A.1 Contextual Statistical Realist Interpretation of Physical Theories

Quantum and classical physical theories describe the properties of the pairs

$$\pi = (\text{physical system}, \text{measuring device}). \tag{A.1}$$

I do not think that understanding of this fact, the *contextual structure of physical theories*, was really Bohr's invention. It was clear to everybody that physical observables are related to properties of physical systems as well as measuring devices. The main invention of N. Bohr was not contextuality, but *complementarity*. Bohr's greatest contribution was the recognition of the fact that there exist complementary experimental arrangements and hence complementary, incompatible, pairs π_1, π_2 of form (A.1). I think nobody can be against the recognition of such a possibility. Why not? Why must all contexts, complexes of physical conditions, be coexisting? Contextuality and complementarity are two well understandable principles (not only of quantum physics, but physics in general).[1]

If we remember about contextuality discussing complementarity we see that complementarity of contexts in quantum physics does not imply complementarity of corresponding objective properties (of elementary particles) contributing into such observables. In particular, contextual complementarity does not imply that elementary particles do not have objective properties at all. In particular, there are no reasons to suppose that it is impossible to provide a kind of hidden variable, HV, description (*ontic description*, see e.g. [167], and section 6.1.2, Chapter 6, of this book) for these objective properties. Mathematically the pair π can be described by two variables λ_s and λ_d representing objective states of a physical system and measuring device. So a physical observable is represented as a function

$$A = A(\lambda_s; \lambda_d). \tag{A.2}$$

On the other hand, an adherent of N. Bohr would argue that "Such a separation and, hence, the description of (properties of) quantum objects and processes themselves (as opposed to certain effects of their interaction with measuring instruments upon latter) are impossible in Bohr's interpretation," [219].

I think that the origin of such an interpretation of complementarity by N. Bohr was the individual interpretation of Heisenberg's uncertainty principle:

$$\Delta q \Delta p \geq \frac{h}{2}. \tag{A.3}$$

Close relation between Bohr's complementarity principle and Heisenberg's uncertainty principle is well known. For two years (1926-1927) N. Bohr could not present any model that could explain both corpuscular behavior (black body radiation, photoelectric effect), and wavelike behavior (two-slit experiment, diffraction) of elementary particles. You can read e.g. in the book [6] how heavy this thinking process was for N. Bohr - really a kind of mental disaster. Only after the derivation by W. Heisenberg of the uncertainty principle, N. Bohr proposed a new model of physical reality based on the principle of *complementarity*. Unfortunately, Heisenberg's uncertainty relation was interpreted as the relation for an individual elementary particle. The main problem was mixing by W. Heisenberg of *individual* and *statistical* uncertainty. For example, in his famous book [108] he discussed

---

[1]It is a pity that Bohr did not use the term contextuality which matches perferctly his discussions on complementarity. He used expressions such as "experimental arrangement".

the uncertainty principle as a relation for an individual system, but derived this principle by using statistical methods![2]

The roots of such individual complementarity can already be found in the first work of W. Heisenberg [107]. At the beginning Heisenberg's quantization procedure was not statistical. It seems, see [107], that W. Heisenberg was sure that he found equations for observables related to individual physical systems. It was a rather common point of view: in classical mechanics the position and momentum of an individual (!) system are described by real numbers, in quantum mechanics – by matrices. W. Heisenberg rightly underlined that the matrix description could not be used for objective position and momentum of electron.

There (as everywhere in XVIII – XX centuries physics) objective reality was, in fact, identified with the continuous (real number) mathematical model of physical reality. Impossibility to create a continuous real number model for motion of electron in Bohr's model of atom was considered by Heisenberg (and many others) as impossibility to assign objective properties to electrons. It was a rather strange passage. But we understand that it was the beginning of XX century and W. Heisenberg used the standard mathematical image of physical reality.[3] However, we would like to remark that even at that time there were attempts to modify continuous mathematical model to reproduce quantum effects, see e.g. Bohr, Kramers, Slater [37] on classical-like quantization based on *difference equations* (instead of differential). In fact, this model stimulated M. Born to introduce the term quantum mechanics. In any case the absence of continuous classical model for motion of electron in Bohr's atom does not imply impossibility to create other, noncontinuous, classical (causal deterministic) models.

Moreover, considerations of W. Heisenberg in [107] even did not imply impossibility to create continuous classical model, although it was claimed by W. Heisenberg and then by N. Bohr. The story is much simpler. First Bohr (with collaborators) tried to create such a model, but was not able; then Heisenberg tried to do the same and did not succeed. Then they

---

[2]We now add the following comment. Recently M. Ozawa [213] discussed this point in very detail and he derived new uncertainty relation. It was also demonstrated experimentally that inequality (A.3) where the quantity Δ is treated as the precision of individual measurement (and not as the standard deviation) can be violated [240].

[3]Comment. Here I wanted to point to the fundamental role played by real numbers mathematics in QM. Borrowing real analysis from classical physics to QM was not so innocent step. By formulating various NO-NO statements about QM one has to remember that these are "real numbers based NO-GOES", cf. with p-adic models in theoretical physics [246], [126].

claimed that such a model did not exist.[4]

W. Heisenberg proposed some mathematical model for some class of observations of position and momentum of electron. These observations satisfy the uncertainty relation. It is not clear why we cannot present other mathematical models for some other observations of position and momentum of electron that would violate this relation? Of course, if we relate Heisenberg's position and momentum to individual electron, then such individualization plays the role of objectification of these quantities. It is rather strange logical circle, but it seems that it was done by W. Heisenberg and N. Bohr. Finally, this objectification in combination with the uncertainty relation implies (for W. Heisenberg and N. Bohr) impossibility to consider other position and momentum variables, distinct from Heisenberg's ones, i.e., represented by another mathematical model. On the other hand, if we use statistical interpretation of uncertainty relation, then there are no reasons for such NO-GO conclusions. We could not prepare statistical ensembles with small dispersions for two variables introduced by W. Heisenberg. But this has nothing to do with impossibility of objectification of these variables.

Unfortunately, many bright scientists used and still use Heisenberg-Bohr's "quantum logic". For instance, the fundamental paper of A. Zeilinger [258] gives us an excelent example of the modern representation of this logic. In principle, A. Zeilinger is looking for a new quantum paradigm. He correctly underlines that the situation in quantum theory, especially large diversity of interpretations, is not so natural. In contrast to, e.g., theory of relativity, there is no quantum analogue of the *principle of relativity*. However, it seems that, for Zeilinger as well as for many other scientists looking for reconsideration of quantum foundations, reconsideration could (and moreover should) be performed as some addition to the orthodox Copenhagen. A. Zeilinger started (as always in this story) with the correct statement that we are dealing with "a quantum phenomenon as the whole entity which comprises both the observed quantum system and the classical measuring apparata." No doubts! The formalism of quantum mechanics (statistical formalism) deals with such a phenomenon. However, then he continued: "It is especially impossible in principle to predict with certainty both through which slit an elementary particle will go and where it will appear in the interference pattern."

We can still interpret this statement in the correct way: the quantum

---

[4]Comment. Thus Bohr and Heisenberg formalized their personal inabilities to go beyond the operational quantum formalism as the fundamental principle of physics.

formalism does not give us such a possibility. Unfortunately, the latter understanding was impossible for orthodox Copenhagenists, since (by unclear reasons) they supposed that quantum theory provided the complete description of physical reality.[5] Zeilinger continued: "I propose that this impossibility to describe the random individual process within quantum mechanics in a complete way is a fundamental limitation of the program of modern science to arrive at a description of the world in every detail." This is the great manifestation of Copenhagen NO-GO.

Another important contribution to quantum foundations that stimulated Bohr to elaborate the complementarity way of thinking about quantum observables was creation of "quantum wave mechanics" by Schrödinger. It is important to recall that E. Schrödinger was sure that he discovered totally new method of quantization [228], "Quantization as the problem for eigenvalues." In his first paper [228] he did not refer to Heisenberg's paper [107]; in the second paper [229] he made a short reference in the sense that W. Heisenberg proposed another method of quantization that was totally different from Schrödinger's one. It is well known that many famous physicists had the great prejudice against Heisenberg's approach to quantization. Schrödinger's wave mechanics was considered by them as the end of quantum mechanics in the spirit of Bohr and Heisenberg; as the possibility to describe "quantum experiments" by using classical theory of partial differential equations (especially strong anti-Heisenberg-Bohr comments were done by Einstein and Wien).

Hence, at that time Heisenberg and Bohr should find some strong arguments in favor of Heisenberg's approach or disappear from the quantum scene. Moreover, the quantum spectacle would end with the trivial final: instead of considering QM as the greatest mystery in the history of science, one would treat it simply as a new part of the well established theory of partial differential equations. In such circumstances N. Bohr found an excellent argument supporting Heisenberg's QM [108], [107] the principle of complementarity with all its NO-GO consequences.

Bohr's complementarity was a kind of *individual complementarity*. Complementary features were regarded to individual physical systems. It is a pity that contextualists N. Bohr and W. Heisenberg related the uncertainty relation not to some special class of measurement procedures of the position and momentum described by quantum formalism, but to the position and momentum of an individual particle. This was the root of

---

[5]The paper of Einstein, Podolsky and Rosen [85] was directed precisely against Heisenberg-Bohr "quantum logic" based on the idea that quantum theory is complete.

the prejudice that the position and momentum even in principle cannot be determined simultaneously.

In fact, the only possible consistent interpretation of Heisenberg's uncertainty principle is the statistical contextual interpretation, see e.g. [196], [197], [25], [26]. It is impossible to prepare such an ensemble of quantum systems that dispersions of both position and momentum observables would be arbitrary small. Everybody would agree that only this statement can be verified experimentally. Contextualism has to be *statistical contextualism* and, consequently, complementarity has to be *statistical contextual complementarity*. Such contextuality and complementarity do not contradict the possibility of creation of a finer description of reality than given by quantum theory.

The complex of experimental physical conditions can be split into two complexes – a preparation procedure and a measurement procedure, see e.g. [48], [26]. A preparation procedure produces a statistical ensemble of physical system. Then a measurement device produces results, values of a physical observable. Nonzero dispersion of this random variable does not imply that an individual physical system does not have objective properties that generate values of the observable via (A.2). Thus the contextual statistical interpretation can be, in principle, extended to the *contextual statistical realist interpretation*.

"Individual complementarity" can be used as an argument against the possibility to create a finer description of physical reality than given by QM. However, statistical complementarity cannot be used as such an argument. By the contextual statistical interpretation it is not forbidden to create such preparation and measurement procedures[6] that position and momentum would be measured with dispersions Δq and Δp such that

$$\Delta q \Delta p < \frac{h}{2}.$$

Of course, such a statement should immediately induce a storm of protests with reference to the principle of complementarity. However, we again recall that the right complementarity principle is contextual and statistical. It is about some special class of measurement and preparation contexts described by the quantum formalism. In particular, we do not consider quantum formalism as a kind of *complete* physical theory.

However, as far as we cannot perform such experiments for quantum systems, it is really impossible to reject Bohr's principle of individual com-

---

[6]Comment. They may be invented as the output of future technological development.

plementarity.[7] What can we do?

We can study consequences of the general statistical contextuality and try to demonstrate that some distinguishing features of quantum theory which are typically associated with individual complementarity, NO-GO complementarity, are, in fact, simple consequences of statistical complementarity, GO-DEEPER complementarity. I did this in a series of papers, see e.g. [141], [142] and Chapter 5. The main consequence of these investigations is that "waves of probability" can be produced in the general situation (including macroscopic systems) due to combination of a few preparation contexts.[8] Thus such "waves" are not directly related to the real wave features of objects participating in experiments. Moreover, my studies demonstrated that in some experiments there can be created other types of probabilistic "waves", namely hyperbolic waves of probability.

In particular, my contextual probabilistic modeling demonstrated that contextual complementarity, wave-particle dualism, is not rigidly coupled to the microworld. Thus we can, in principle, perform experiments with macro systems that would demonstrate "wave-particle duality".

For example, my student Ja. Volovich performed a numerical experiment, a classical analogue of the well known quantum two-slit experiment, see [136]. Charged particles are scattered on the flat screen with two slits and hit the second screen. It was shown that the probability distribution on the second screen when both slits are open is not given as the sum of the distributions for each slit separately, but has an extra interference term expressing the quantum rule of addition of probabilistic alternatives. We have two theoretical descriptions of this experiment:

- the quantum-like statistical description;
- the Newtonian classical description.

Both theories predict the same statistical distribution of spots on the registration screen. Quantum-like theory operates with complex waves of probability; there is uncertainty, Heisenberg-like, relation for position and momentum. Of course, this relation is the statistical one.

Suppose now that some observer cannot approach the level of the Newtonian description, e.g., such an observer is a star-size observer and his mea-

---

[7]We may try to produce various realistic models underlying quantum formalism, for example, Bohmian mechanics, stochastic electrodynamics. However, it seems that all such models would be more or less automatically rejected by the quantum community.

[8]In the two-slit experiment we consider the screen with slits as a part of the ensemble-preparation device and the screen with photoemulsion as the measurement device.

surement devices produce non-negligible perturbations of our macroscopic charged balls. This observer might speculate about the impossibility to create the objective phase-space description and even about wave-features of macroscopic charged balls. This experiment might be used as an argument against the orthodox Copenhagen NO-GO in experiments.

We recognize that at the moment there is one (and seems just one) argument supporting individual complementarity, namely Bell's inequality.[9] However, I have great doubts that the experimental violation of Bell's inequality can be interpreted as an argument against the possibility of creation of the HV ontic description (or even against realism) or locality, see e.g. my works [133], [137] (see also papers of L. Accardi, L. Ballentine, W. De Myunck in [139]; see also [140] on the contextual statistical realist interpretation of the GHZ-paradox).

Finally, we remark that the possibility of (A.2)-description implies that "quantum randomness" does not differ from "classical randomness". Of course, this contradicts to the orthodox quantum view on randomness as *fundamental or irreducible randomness*. Instead of fundamental irreducible quantum randomness, I prefer to consider well understandable theory of context dependent probabilities.

We propose a new fundamental principle – analog of the basic principle of relativity theory (see also Chapter 5 and paper [152]):

The principle of contextual relativity of probabilities.[10] *Contextuality means that all probabilities depend on complexes of physical conditions, contexts:* $P(E) \equiv P_C(E)$*, where C is experimental context. It is meaningless to speak about probability without determining a complex of physical conditions, context.*

We remark that the conventional probability theory based on Kolmogorov measure theoretical axiomatics [177], is not contextual. In Kolmogorov's theory the probability space can be fixed once and for all. We need not remember that probabilities depend on complexes of physical conditions and can use just the symbol of abstract probability $\mathbf{P}$.[11]

---

[9]Some people use this framework to support quantum nonlocality.

[10]Comment. Unfortunately, this principle does not specify quantum theory completely. Contextuality is a more general notion than quantumness. QM is contextual, but not any contextual probability model is quantum. The main problem of modern quantum foundations is to specify additional principles which in combination with contextuality would characterize QM. In section 5.4, Chapter 5, we made some step in this direction. We also recall similar efforts of QBists, sections 6.7.3 and 6.7.5, Chapter 6.

[11]We remark that A. Kolmogorov understood well contextual dependence of probabil-

A.2 Citation with Comments

In this section we shall present some citations on orthodox quantum theory and our contextual statistical realist comments. We use, in particular, collections of Bohr's views presented in papers of H. Folse and A. Plotnitsky, see [91], [219] (see also [92], [220] - [222]).

(S1) "In contrast to ordinary mechanics, the new quantum mechanics does not deal with a space-time description of the motion of atomic particles. It operates with manifolds of quantities which replace the harmonic oscillating components of the motion and symbolize the possibilities of transition between stationary states in conformity with the correspondence principle", N. Bohr.

This is simply the recognition of the restrictiveness of the domain of applications of quantum theory. I would like to interpret this as the recognition of incompleteness of quantum theory. However, it was not so for N. Bohr:

(S2) " ... the quantum postulate implies that any observation of atomic phenomena will involve an interaction with the agency of observation not to be neglected. Accordingly, an independent reality in the ordinary physical sense can neither be ascribed to the phenomena nor to the agencies of observation," N. Bohr.

The first part of this citation is the manifestation of contextuality. However, I cannot understand what kind of logic N. Bohr used to proceed to the second part. The second part can be interpreted as the declaration of the impossibility of the objective, ontic description of reality.

(S3) "... to reserve the word phenomenon for comprehension of effects observed under given experimental conditions... These conditions, which include the account of the properties and manipulation of all measuring instruments essentially concerned, constitute in fact the only basis for the definition of the concepts by which the phenomenon is described," N. Bohr.

I would agree if the last sentence would be continued as "is described in quantum formalism."

(S4) "...by the very nature of the situation prevented from differentiat-

---

ities, see section 2 of his book [177] devoted to experimental applications of his theory. However, this contextuality was not present in his mathematical formalism. It was the terrible mistake of Kolmogorov. In fact, he tried to improve the situation and published paper, see references in [156], in that he noticed that probabilities are contextual probabilities and even used corresponding symbol. However, the paper was published only in Russian and even in Soviet Union it was forgotten. Recently the pupil of Kolmogorov, A. Shiryaev, brought my attention to this paper.

ing sharply between an independent behavior of atomic objects and their interaction with the means of observation indispensable for the definition of the phenomena," N. Bohr.

I would agree if the last sentence would be continued as "of the phenomena described by quantum formalism."[12]

(S5) "There are two forms in which quantum mechanics may be expressed, based on Heisenberg's matrices and Schrödinger's wave function, respectively. The second of these is not connected directly with classical mechanics. The first is in close analogy with classical mechanics, as it may be obtained from classical mechanics simply by making the variables of classical mechanics into noncommuting quantities satisfying the correct commutation relations." P. Dirac [78].

A.3   On Romantic Interpretation of Quantum Mechanics

Finally, we ask: "Why the realist interpretation is not so popular in quantum community?"

The common opinion: this is the direct consequence of experiments with elementary particles. Well, I do not think so. Of course, interference experiments with massive particles should induce reconsideration of methods of classical statistical mechanics. As well as discreteness of energy levels should induce reconsideration of classical continuous real model of physics. However, such reconsiderations would merely be mathematical. And it seems (at least for me) that they were merely mathematical. Observables take only discrete values – consider 'discrete' number systems instead of 'continuous'. The standard probability calculus (created for one fixed sample space prepared under stable physical conditions) does not work – create the new one. And this was very successfully done. However, there were no reasons to create new quantum philosophy, based on Bohr's principle of complementarity and the individual interpretation of Heisenberg's uncertainty relations. Nevertheless, such a new philosophy was invented (merely by N. Bohr) and, moreover, it was recognized as philosophy of

---

[12]Comment. My main message is that one should distinguish a mathematical model from real physical phenomena. QM is just a special mathematical model of micro-phenomena. However, it seems that the fathers of QM identified the basic features of this mathematical model with physical reality. (We recall that, although they rejected the possibility of the realist interpretation of QM, they did not reject reality!) I want to say that in future other models of micro-phenomena may be elaborated. Their features may differ essentially from features of QM, cf., e.g., with the random field modeling of micro-phenomena (see section 5.2.2, Chapter 5).

modern physics. Since this recognition, all realist models for experiments with quantum systems were more or less automatically rejected, see, e.g,. A. Lande's statistical contextual realist model for diffraction [193]. Lande's model looks quite natural; here we need not apply to wave-particle dualism, collapse and so on... It was simply rejected. Bohmian mechanics, see e.g. [36], – well, it has its disadvantages, but merely, mathematical.

Finally, typically Bell's inequality arguments were interpreted as they should be interpreted in the orthodox quantum framework, despite very strong counter-arguments. If all these counter-arguments were taken into account, Bell's theoretical reasoning would not be totally justified or the present interpretation of the corresponding experiments.

I suspect that the main reason for this rather strange situation in modern physics is the great attraction of romantic spirit of orthodox quantum philosophy. All these nonreal things, wave-particle dualism, collapse, nonlocality, irreducible randomness, were attractive for some of creators as well as further quantum generations. These are different stories to discuss merely mathematical modification of real-continuous model of classical statistical mechanics or to declare scientific revolution. Thus the orthodox Copenhagen interpretation can be considered as a *romantic stage* of development of physics. It is not easy to argue against romanticism. (It is very attractive!!!) Probably we need not do this. I hope that realism (as the history of literature shows us) would sooner or later come to physics.

We underline that our analysis of views of N. Bohr and H. Heisenberg demonstrated that by using the statistical interpretation of the principle of complementarity and uncertainty relations we escape the contradiction between the quantum formalism and realism. Statistical approach to Bohr's contextualism does not contradict the possibility to construct a realist pre quantum model.

Appendix B

Analogy between non-Kolmogorovian Probability and non-Euclidean Geometry

The important theme of this book is about a possibility to "embed" a noclassical probabilisty models, e.g., quantum probability, into the classical (Kolmogorov) probability model (Chapters 5 and 8). From the viewpoint of the analogy geometry-probability, it would not be surprising if a non-Kolmogorovian model were embedded in some (may be nonunique) way into the Kolmogorovian model. We know that non-Euclidean geometries can be *modeled* by using surfaces in the Euclidean space. As was shown in Chapter 8, in this aspect probability is similar to geometry; it is really possible to embed, e.g., quantum probability into the classical probability model. The embedding proposed in Chapter 8 is based on treatment of quantum probabilities (correlations) as conditional classical probabilities (correlations). The EPR-Bohm correlations used in Bell's "no-go theorem" can be treated in this way.

We know that a non-Euclidean geometry having very exotic features (from the Euclidean viewpoint) can be modeled by using surfaces in the Euclidean space. Let us turn to the *Lobachevskian geometry,* consider, for example, the so called hyperbolic plane. There are various models of this plane based on the Euclidean plane. Consider, for example, the Poincaré disc model. It is based on the interior of a circle (in the Euclidean plane) and lines are represented by arcs of circles that are orthogonal to the boundary circle, plus diameters of the boundary circle. We stress that, although in this representation the hyperbolic plane is given by a domain of the Euclidean plane, the correspondence between the basic entities of the models is nontrivial. Lines of the hyperbolic geometry are not at all straight lines of the Euclidean geometry. Therefore if we hope to embed the quantum model of probability into the classical model of probability. And we have seen that quantum probability model can be embedded into classical probability model only in a nontrivial way: we cannot proceed with unconditional (probabilistic) compatibility. Embedding is possible only by using conditional compatibility of observables.

Bibliography

[1] Acacio de Barros, J. and Suppes, P. (2009). Quantum mechanics, interference, and the brain, *J. Math. Psych.* **53**, pp. 306–313.

[2] Accardi, L. (1997). *Urne e Camaleoni: Dialogo sulla Realta, le Leggi del Caso e la Teoria Quantistica* (Il Saggiatore, Rome).

[3] Accardi, L. (2004). *Dialogs about Quantum Mechanics, R&S Dynamics*, Moscow, (in Russian).

[4] Accardi, L. (2015). *Dialogs about quantum mechanics*, (in Japanese).

[5] Accardi, L., Khrennikov, A., Ohya, M., Tanaka, Y. and Yamao, I. Non-Kolmogorovian probabilities and its application to unconscious inference in visual perception processes, *Open Systems and Information Dynamics*, to be published.

[6] Adenier, G. and Khrennikov, A. (2007). Is the fair sampling assumption supported by EPR experiments?, *J. Phys. B: Atomic, Molecular Opt. Phys.* **40**, pp. 131–141.

[7] Albeverio, S., Khrennikov, A. and Kloeden, P. (1999). Memory retrieval as a p-adic dynamical system, *Biosystems* **49**, pp. 105–115.

[8] Arndt, M., Juffmann, Th. and Vedral, V. (2009). Quantum physics meets biology, *HFSP J.* **3**, 6, pp. 386–400.

[9] Asano, M., Ohya, M., Tanaka, Y., Khrennikov, A. and Basieva, I. (2010). On application of Gorini-Kossakowski-Sudarshan-Lindblad equation in cognitive psychology, *Open Systems and Information Dynamics* **17**, pp. 1–15.

[10] Asano, M., Ohya, M., Tanaka, Y., Basieva, I., Khrennikov, A. (2011). Quantum-like model of brain's functioning: Decision making from decoherence, *J. Theor. Biology* **281**, pp. 56–64.

[11] Asano, M., Basieva, I., Khrennikov, A., Ohya, M. and Yamato, I. (2013). Non-Kolmogorovian approach to the context-dependent systems breaking the classical probability law, *Found. Phys.* **43**, pp. 895–911.

[12] Asano, M., Basieva, I., Khrennikov, A., Ohya, M. and Tanaka, Yo. (2012). Quantum-like generalization of the Bayesian updating scheme for objective and subjective mental uncertainties, *J. Math. Psych.* **56**, pp. 166–175.

[13] Asano, M., Basieva, I., Khrennikov, A., Ohya, M., Tanaka, Y. and Yamato, I. (2012). Quantum-like model for the adaptive dynamics of the genetic

regulation of E. coli's metabolism of glucose/lactose, *Syst. and Synth. Biol.* **6**, pp. 1–7.

[14] Asano, M., Basieva, I., Khrennikov, A., Ohya, M., Tanaka, Y. and Yamato, I. (2013). A model of epigenetic evolution based on theory of open quantum systems, *Syst. and Synth. Biol.* **7**, pp. 161–173.

[15] Asano, M., Khrennikov, A., Ohya, M., Tanaka, Y. and Yamato, I. (2014). Violation of contextual generalization of the LeggettGarg inequality for recognition of ambiguous figures, *Phys. Scr.* **T163**, 014006.

[16] Asano, M., Khrennikov, A., Ohya, M., Tanaka, Y. and Yamato, I. (2015). *Quantum Adaptivity in Biology: From Genetics to Cognition* (Springer, Heidelberg-Berlin-New York).

[17] Aspect, A. (1983). *Three Experimental Tests of Bell Inequalities by the Measurement of Polarization Correlations between Photons* (Orsay Press, Orsay).

[18] Aspect, A., Dalibard, J. and Roger, G. (1982). Experimental test of Bell's inequalities using time-varying analyzers, *Phys. Rev. Lett.* **49**, 1804.

[19] Aspect, A. Comment on "A classical model of the EPR experiment with quantum mechanical correlations and Bell inequalities", in G. T. Moore and M. O. Sculy (eds), *Frontiers of Nonequilbrium Statistical Physics* (Plenum Press, New York, London), pp. 185–189.

[20] Atmanspacher, H., Bishop, R. C. and Amann, A. (2001). Extrinsic and intrinsic irreversibility in probabilistic dynamical laws, in A. Khrennikov (ed), *Foundations of Probability and Physics* (World Scientific, Singapore), pp. 50–70.

[21] Atmanspacher, H. (2002). Determinism is Ontic, Determinability Is Epistemic, in H. Atmanspacher and R. C. Bishop (eds.), *Between Chance and Choice: Interdisciplinary Perspectives on Determinism* (Thorverton UK, Imprint Academic), pp. 49–74.

[22] Atmanspacher, H. and Filk, Th. (2012). Temporal nonlocality in bistable perception, in A. Khrennikov, H. Atmanspacher, A. Migdall and S. Polyakov (eds.), *Quantum Theory: Reconsiderations of Foundations - 6, Special Section: Quantum-like decision making: from biology to behavioral economics*, (AIP Conf. Proc.) **1508**, pp. 79–88.

[23] Auyang, S. Y. (1995). *How is Quantum Field Theory Possible?* (Oxford University Press, Oxford).

[24] Avis, D., Fischer, P., Hilbert, A. and Khrennikov, A. (2009). Single, Complete, Probability Spaces Consistent With EPR-Bohm-Bell Experimental Data, *Foundations of Probability and Physics-5* **750** (AIP, Melville, NY), pp. 294–301.

[25] Ballentine, L. E. (1970). The statistical interpretation of quantum mechanics, *Rev. Mod. Phys.* **42**, pp. 358–381.

[26] Ballentine, L. E. (1989). *Quantum Mechanics* (Englewood Cliffs, New Jersey).

[27] Ballentine, L. E. (1990). Limitations of the projection postulate, *Found. Phys.* **20**, pp. 1329–1990.

[28] Basieva, I., Khrennikov, A., Ohya, M. and Yamato, I. (2011), Quantum-like

interference effect in gene expression: glucose-lactose destructive interference, *Syst. Synth. Biol.* **5**, pp. 59–68.

[29] Bell, J. S. (1964). On the Einstein-Podolsky-Rosen paradox, *Physics 1* **1**, pp. 195–200.

[30] Bell, J. S. (1982). On the impossible pilot wave, *Ref.TH.3315-CERN*, p. 15.

[31] Bell, J. (1987). *Speakable and Unspeakable in Quantum Mechanics* (Cambridge Univ. Press, Cambridge).

[32] Beltrametti, E. and Cassinelli, G. (1979). Properties of states in quantum logics, in: G. T. de Francia (ed.), *Problems in the Frontiers of Physics* (North Holland).

[33] Bernardo, J. M. and Smith, A. F. M. (1994). *Bayesian Theory* (Wiley, Chichester).

[34] Bernstein, S. N. (1927). *Theory of Probability* (Gosizd, Moscow), in Russian.

[35] Birkhoff, J. and von Neumann, J. (1936). The logic of quantum mechanics, *Annals of Mathematics* **37**, N 4, pp. 823–843.

[36] Bohm, D. and Hiley, B. (1993). *The Undivided Universe: An Ontological Interpretation of Quantum Mechanics* (Routledge and Kegan Paul, London).

[37] Bohr, N., Kramers, H. A., Slater, J. C. (1924). The quantum theory of radiation, *Philosophical Magazine* **47**, pp. 785–802.

[38] Bohr, N. (1935). Can quantum-mechanical description of physical reality be considered complete?, *Phys. Rev.* **48**, pp. 696–702.

[39] Bohr, N. (1938). The causality problem in atomic physics, in J. Faye and H. J. Folse (eds.), (1987). *The Philosophical Writings of Niels Bohr, Volume 4: Causality and Complementarity, Supplementary Papers* (Ox Bow Press; Woodbridge, CT), pp. 94–121.

[40] Bohr, N. (1987). *The Philosophical Writings of Niels Bohr*, 3 vols. (Ox Bow Press, Woodbridge, CT).

[41] Boole, G. (1958). *An Investigation of the Laws of Thought* (Dover Edition, New York).

[42] Boole, G. (1862). On the theory of probabilities, *Phil. Trans. Royal Soc. London* **152**, pp. 225–242.

[43] Brukner, C. and Zeilinger, A. (1999). Malus' law and quantum information, *Acta Physica Slovava* **49** 4, pp. 647–652.

[44] Brukner, C. and Zeilinger, A. (1999). Operationally invariant information in quantum mechanics, *Phys. Rev. Lett.* **83** 17, pp. 3354–3357.

[45] Brukner, C. and Zeilinger, A. (2009). Information invariance and quantum probabilities, *Found. Phys.* **39**, p. 677–686.

[46] Brukner, C., On the quantum measurement problem, Preprint arXiv:1507.05255 [quant-ph].

[47] Buscemi, F., D' Ariano, G. M. and Perinotti, P. (2004). There exist nonorthogonal quantum measurements that are perfectly repeatable, *Phys. Rev. Lett.* **92**, 070403-1–070403-4.

[48] Busch, P., Grabowski, M. and Lahti, P. (1995). *Operational Quantum*

Physics (Springer Verlag, Berlin).

[49] Busemeyer, J. B., Wang, Z. and Townsend, J. T. (2006). Quantum dynamics of human decision making, *J. Math. Psychology* **50**, pp. 220–241.

[50] Busemeyer, J. R. and Bruza, P. D. (2012). *Quantum Models of Cognition and Decision* (Cambridge Univ. Press, Cambridge).

[51] Busemeyer, J. R., Wang, Z., Khrennikov, A. and Basieva, I. (2014). Applying quantum principles to psychology, *Phys. Scr.* **T163**, 014007.

[52] Caves, C. M., Fuchs, C. A. and Schack, R. (2002). Quantum probabilities as Bayesian probabilities, *Phys. Rev. A* **65**, 022305.

[53] Caves, C. M., Fuchs, C. A. and Schack, R. (2007). Subjective probability and quantum certainty, *Stud. Hist. Phil. Mod. Phys.* **38**, pp. 255–274.

[54] Chaitin, G. J. (1969). On the simplicity and speed of programs for computing infinite sets of natural numbers, *Journal of the ACM* **16**, pp. 407–422.

[55] Chaitin, G. J. (2005). *Meta Math!* (Pantheon Books, New York).

[56] Chalmers, D. J. (1995). Facing up to the hard problem of consciousness, *J. Cons. Studies* **2**(3), pp. 200–219.

[57] Chiribella, G., D'Ariano, G. M. and Perinotti, P. (2010). Probabilistic theories with purification, *Phys. Rev. A* **81**, 062348.

[58] Chiribella, G., D'Ariano, G. M. and Perinotti, P. (2012). Informational axioms for quantum theory, in *Foundations of Probability and Physics - 6, AIP Conf. Proc.* **1424**, pp. 270–279.

[59] Christensen, B. G., Mc Cusker, K. T., Altepeter, J., Calkins, B., Gerrits, T., Lita, A., Miller, A., Shalm, L. K., Zhang, Y., Nam, S. W., Brunner, N., Lim, C. C. W., Gisin, N. and Kwiat, P. G. (2013). Detection-loophole-free test of quantum nonlocality, and applications, *Phys. Rev. Lett.* **111**, pp. 1304–1306.

[60] Church, A. (1940). On the concept of a random sequence, *Bull. Amer. Math. Soc.* **46**, pp. 130–135.

[61] Conte, E., Todarello, O., Federici, A., Vitiello, F., Lopane, M., Khrennikov, A. and Zbilut, J. P. (2006). Some remarks on an experiment suggesting quantum-like behavior of cognitive entities and formulation of an abstract quantum mechanical formalism to describe cognitive entity and its dynamics, *Chaos, Solitons and Fractals* **31**, pp. 1076–1088.

[62] Conte, E., Khrennikov, A., Todarello, O., Federici, A., Mendolicchio, L. and Zbilut, J. P. (2008). A preliminary experimental verification on the possibility of Bell inequality violation in mental states, *Neuroquantology* **6**, pp. 214–221.

[63] D'Ariano, G. M. (2007). Operational axioms for quantum mechanics, in Adenier et al. (eds.), *Foundations of Probability and Physics-3* (AIP Conf. Proc.) **889**, pp. 79–105.

[64] D'Ariano, G. M. (2011). Physics as information processing, in *Advances in Quantum Theory*, (AIP Conf. Proc.) **1327**, pp. 7–16.

[65] Dasgupta, A. (2011). Mathematical foundations of randomness, *Philosophy of Statistics. Handbook of Philosophy of Science*, Vol. 7 (North-Holland), pp. 641–710.

[66] Davies, E. and Lewis, J. (1970). An operational approach to quantum prob-

ability, *Comm. Math. Phys.* **17**, pp. 239–260.

[67] De Broglie, L. (1964). *The Current Interpretation of Wave Mechanics: a Critical Study* (with a chapter by A. E. Silva), (Elsevier).

[68] De Finetti, B. (1931). Probabilismo, *Logos* **14**, p. 163; transl. (1989). Probabilism, *Erkenntnis* **31**, pp. 169–223.

[69] De Finetti, B. (2008). *Philosophical Lectures on Probability* (Springer Verlag, Berlin-New York).

[70] de Laplace, P.-S. (1819, 1952). *A Philosphical Essay on Probabilities* (Dover), translated from 6th French edn.

[71] De Muynck, W. M. (2002). *Foundations of Quantum Mechanics, an Empiricists Approach.* (Kluwer, Dordrecht).

[72] De Raedt, K., Keimpema, K., De Raedt, H., Michielsen, K. and Miyashita, S. (2006). A local realist model for correlations of the singlet state, *The European Physical Journal B* **53**, pp. 139–142.

[73] De Raedt, H., De Raedt, K., Michielsen, K., Keimpema, K. and Miyashita, S. (2007). Event-based computer simulation model of Aspect-type experiments strictly satisfying Einstein's locality conditions, *Phys. Soc. Japan* **76**, 104005.

[74] De Raedt, H., Jin, F. and Michielsen, K. (2013). Data analysis of Einstein-Podolsky-Rosen-Bohm laboratory experiments, The Nature of Light: What are Photons?, in C. Roychoudhuri, H. De Raedt and A. F. Kracklauer (eds.), *Proc. of SPIE*, Vol. 8832, 88321.

[75] Diaconis, P., Holmes, S. and Montgomery, R. (2007). Dynamical bias in the coin toss, *SIAM Rev.* **49**, p. 211.

[76] Dirac, P. A. M. (1927). The physical interpretation of the quantum dynamics, *Proc. Royal Soc. London A* **113**, p. 621.

[77] Dirac, P. A. M. (1942). The physical interpretation of quantum mechanics, *Proc. Roy. Soc. London A* **180**, pp. 1–39.

[78] Dirac, P. (1945). On the analogy between classical and quantum mechanics, *Rev. Mod. Phys.* **17**, pp. 195–199.

[79] Dirac, P. A. M. (1958). *The Principles of Quantum Mechanics* (Oxford Clarendon), reprint (1995).

[80] Doob, J. L. (1994). The development of rigor in mathematical probability, in J. P. Pier (ed.), *Development of Mathematics 1900–1950* (Birkhäuser, Basel, Boston, Berlin), pp. 157–170.

[81] Dragovich, B., Khrennikov, A., Kozyrev, S. V. and Volovich, I. V. (2009). On *p*-adic mathematical physics, *P-Adic Numbers, Ultrametric Analysis, and Applications* **1**, N 1, pp. 1–17.

[82] Dubischar, D., Gundlach, V. M., Steinkamp, O. and Khrennikov, A. (1999). A *p*-adic model for the process of thinking disturbed by physiological and information noise, *J. Theor. Biology* **197**, pp. 451–467.

[83] Dzhafarov, E. N. and Kujala, J. V. (2012). Selectivity in probabilistic causality: Where psychology runs into quantum physics, *J. Math. Psych.* **56**, pp. 54–63.

[84] Dzhafarov, E. N. and Kujala, J. V. (2014). On selective influences, marginal selectivity, and Bell/CHSH inequalities, *Topics in Cognitive Science* **6**, pp.

121–128.

[85] Einstein, A., Podolsky, B. and Rosen, N. (1983). Can quantum-mechanical description of physical reality be considered complete?, in: J. A. Wheeler, W. H. Zurek, (eds.), *Quantum Theory and Measurement* (Princeton University Press, Princeton NJ), pp. 138–141.

[86] Einstein, A. and Infeld, L. (1961). *Evolution of Physics: The Growth of Ideas from Early Concepts to Relativity and Quanta* (Simon and Schuster, New-York).

[87] Feller, W. (1968). *An Introduction to Probability Theory and its Applications*, Vol. 1 (John Wiley & Sons, New York).

[88] Feynman, R. and Hibbs, A. (1965). *Quantum Mechanics and Path Integrals* (McGraw-Hill, New York).

[89] Feynman, R. P. (1987). Negative probability, in: B. J. Hiley and F. D. (eds.), *Quantum Implications, Essays in Honour of David Bohm* (Routledge and Kegan Paul, London), pp. 235–246.

[90] Fichtner, K. H., Fichtner, L., Freudenberg, W. and Ohya, M. (2008). On a quantum model of the recognition process, *QP-PQ: Quantum Prob. White Noise Analysis* **21**, World Scientific, pp. 64–84.

[91] Folse, H. (2001). *Bohr's best bits*, preprint.

[92] Folse, H. (2014). The methodological lesson of complementarity: Bohr's naturalistic epistemology, *Phys. Scr.* **T163**, 014001.

[93] Fuchs, C. A. (2002). Quantum mechanics as quantum information (and only a little more), in A. Khrennikov (ed.), *Quantum Theory: Reconsideration of Foundations, Ser. Math. Modeling* **2** (Växjö University Press, Växjö), pp. 463–543.

[94] Fuchs, C. A. (2002). The anti-Växjö interpretation of quantum mechanics, *Quantum Theory: Reconsideration of Foundations, Ser. Math. Model.* **2** (Växjö University Press, Växjö), pp. 99–116.

[95] Fuchs, C. (2007). Delirium quantum (or, where I will take quantum mechanics if it will let me), in G. Adenier, C. Fuchs and A. Yu. Khrennikov (eds.), Foundations of Probability and Physics-3, *Ser. Conference Proceedings* **889** (American Institute of Physics, Melville, NY), pp. 438–462.

[96] Fuchs, C. A. and Schack, R. (2011). A quantum-Bayesian route to quantum-state space, *Found. Phys.* **41**, pp. 345–356.

[97] Fuchs, C. A. and Schack, R. (2013). Quantum-Bayesian Coherence, *Rev. Mod. Phys.* **85**, p. 1693.

[98] Fuchs, C. A. and Schack, R. (2014). QBism and the Greeks: why a quantum state does not represent an element of physical reality, *Phys. Scr.* **90**, 015104.

[99] Fuchs, C. A., Mermin, N. D. and Schack, R. (2014). An introduction to QBism with an application to the locality of quantum mechanics, *Am. J. Phys.* **82**, pp. 749–754.

[100] Gillies, D. (2011). *An Objective Theory of Probability* (Routledge Revivals, Abingdon).

[101] Giustina, M., Mech, Al., Ramelow, S., Wittmann, B., Kofler, J., Beyer, J., Lita, A., Calkins, B., Gerrits, Th., Woo Nam, S., Ursin, R. and Zeilinger,

A. (2013). Bell violation using entangled photons without the fair-sampling assumption, *Nature* **497**, pp. 227–230.

[102] Giustina, M. et al., (2015). A significant-loophole-free test of Bell's theorem with entangled photons, arXiv:1511.03190.

[103] Gnedenko, B. V. (1998). *Theory of Probability* (CRC Press).

[104] Griffiths, R. B. (2002). *Consistent Quantum Theory* (Cambridge Univ. Press, Cambridge).

[105] Hameroff, S. (1994). Quantum coherence in microtubules. A neural basis for emergent consciousness?, *J. of Cons. Stud.* **1**, pp. 91–118.

[106] Haven, E. and Khrennikov, A. (2012). *Quantum Social Science* (Cambridge Univ. Press, Cambridge).

[107] Heisenberg, W. (1925). Über quantentheoretische Umdeutung kinematischer und mechanischer Beziehungen, *Zeits. für Physik* **33**, p. 879. English translation: (1968). Quantum theoretical re-interpretation of kinematic and mechanical relations in B. L. van der Waerden (trans., ed.), *Sources of Quantum Mechanics* (New York: Dover), pp. 261–276.

[108] Heisenberg, W. (1930). *Physical Principles of Quantum Theory* (Chicago Univ. Press, Chicago).

[109] Hensen, B. et al. (2015). Loophole-free Bell inequality violation using electron spins separated by 1.3 kilometers, *Nature*, doi: 10.1038/nature.15759.

[110] Hess, K. and Philipp, W. (2003). Exclusion of time in Mermin's proof of Bell-type inequalities, in A. Yu. Khrennikov (ed.), Quantum Theory: Reconsideration of Foundations-2, *Ser. Math. Model.* **10** (Växjö University Press, Växjö), pp. 243–254.

[111] Hess, K. and Philipp, W. (2005). Bell's theorem: critique of proofs with and without inequalities, in G. Adenier, A. Yu. Khrennikov (eds.), Foundations of Probability and Physics-3, *Ser. Conference Proceedings* **750** (American Institute of Physics, Melville, NY), pp. 150–155.

[112] Hess, K. (2007). In memoriam Walter Philipp, in G. Adenier, C. Fuchs and A. Yu. Khrennikov (eds.), Foundations of Probability and Physics-3, *Ser. Conference Proceedings* **889** (American Institute of Physics, Melville, NY), pp. 3–6.

[113] Hess, K. (2014). *Einstein was right!* (Pan Stanford Publ., Singapore).

[114] Hilbert, D. (1900). *Mathematische Probleme* (Göttinger Nachrichten), pp. 253–297.

[115] Hilbert, D. (1901). Mathematische Probleme, *Archiv der Mathematik und Physik*, 3d ser. **1**, pp. 44–63, pp. 213–237.

[116] Hilbert, D. (1902). Mathematical Problems, *Bull. American Math. Soc.* **8**, pp. 437–479.

[117] Holevo, A. S. (1982). *Probabilistic and Statistical Aspects of Quantum Theory* (North-Holland, Amsterdam).

[118] Holevo, A. S. (2001). *Statistical Structure of Quantum Theory* (Springer, Berlin-Heidelberg).

[119] Jaeger, G. (2013). *Quantum Objects: Non-Local Correlation, Causality and Objective Indefiniteness in the Quantum World* (Springer, Berlin-New York).

[120] James, W. (1884). The dilemma of determinism, in (2009). *The Will to Believe and Other Essays in Popular Philosophy* (The Project Gutenberg Ebook), www.gutenberg.net, pp. 145–183.

[121] James, W. (1882). On some Hegelisms, in (2009). *The Will to Believe and Other Essays in Popular Philosophy* (The Project Gutenberg Ebook), www.gutenberg.net, pp. 263–298.

[122] James, W. (1890). *The Principles of Psychology* (Henry Holt and Co., New York), Reprinted 1983 (Harvard Univ. Press, Boston).

[123] Khrennikov, A. (1985). The Feynman measure on the phase space and symbols of infinite-dimensional pseudo-differential operators, *Math. Notes* **37**, N 5, pp. 734–742.

[124] Khrennikov, A. (1987). Infinite-dimensional pseudo-differential operators, *Izvestia Akademii Nauk USSR, ser. Math.* **51**, N 6, pp. 46–68.

[125] Khrennikov, A. (2015). Quantum-like modeling of cognition, *Frontiers in Physics* **3**, 00077.

[126] Khrennikov, A. Yu. (1994). *P-adic Valued Distributions and Their Applications to the Mathematical Physics* (Kluwer, Dordreht).

[127] Khrennikov, A. (1997). *Non-Archimedean analysis: quantum paradoxes, dynamical systems and biological models* (Kluwer, Dordreht).

[128] Khrennikov, A. (1998). Human subconscious as the p-adic dynamical system, *J. Theor. Biology* **193**, pp.179–196.

[129] Khrennikov, A. (1999). p-adic information spaces, infinitely small probabilities and anomalous phenomena, *J. of Scientific Exploration* **4**, N 13, pp. 665–680.

[130] Khrennikov, A. (1999). Classical and quantum mechanics on information spaces with applications to cognitive, psychological, social and anomalous phenomena, *Found. of Physics* **29**, N 7, pp. 1065–1098.

[131] Khrennikov, A. (1999). Description of the operation of the human subconscious by means of p-adic dynamical systems, *Dokl. Akad. Nauk* **365**, N 4, pp. 458–460.

[132] Khrennikov, A. (1997). *Superanalysis* (Nauka, Fizmatlit, Moscow), in Russian. English translation: (1999). Kluwer, Dordrecht.

[133] Khrennikov, A. Yu. (1999). *Interpretations of Probability* (VSP Int. Sc. Publishers, Utrecht/Tokyo).

[134] Khrennikov, A. (2000). Non-Kolmogorov probability models and modified Bell's inequality, *J. Math. Phys.* **41**, N 4, pp. 1768–1777.

[135] Khrennikov, A. (1999). Statistical measure of ensemble nonreproducibility and correction to Bell's inequality, *Il Nuovo Cimento* **B 115**, N 2, pp. 179–184.

[136] Khrennikov, A. Yu., Volovich, I. (2000). *Numerical experiment on interference for macroscopic particles*, Preprint: quant-ph/0111159.

[137] Khrennikov, A. Yu. (2000). A perturbation of CHSH inequality induced by fluctuations of ensemble distributions, *J. Math. Physics* **41**, N 9, pp. 5934–5944.

[138] Khrennikov, A. (2000). Classical and quantum mechanics on p-adic trees of ideas, *BioSystems* **56**, pp. 95–120.

[139] A. Khrennikov (ed.) (2001). *Foundations of Probability and Physics* (World Scientific, Singapore).

[140] Khrennikov, A. Yu. (2001). Contextualist viewpoint to Greenberger-Horne-Zeilinger paradox, *Phys. Lett. A* **278**, pp. 307–314.

[141] Khrennikov, A. Yu. (2001). Linear representations of probabilistic transformations induced by context transitions, *J. Phys. A: Math. Gen.* **34**, pp. 9965–9981.

[142] Khrennikov, A. Yu. (2001). Origin of quantum probabilities, in A. Khrennikov (ed.), *Foundations of Probability and Physics* (Växjö-2000, Sweden; World Scientific, Singapore), pp. 180–200.

[143] Khrennikov, A. (2002). Växjö interpretation of quantum mechanics, in *Quantum Theory: Reconsideration of Foundations, Ser. Math. Modelling* (Växjö Univ. Press), Vol. 2, pp. 163–170; Preprint: arXiv:quant-ph/0202107.

[144] Khrennikov, A. Yu. and Volovich, I. V. (2002). Local Realism, Contextualism and Loopholes in Bell's Experiments, in *Proc. Conf. Foundations of Probability and Physics-2, Ser. Math. Modelling* **5** (Växjö Univ. Press), pp. 325–344.

[145] Khrennikov, A. (2003). Quantum-like formalism for cognitive measurements, *Biosystems* **70**, pp. 211–233.

[146] Khrennikov, A. (2004). Contextual approach to quantum mechanics and the theory of the fundamental prespace, *J. Math. Phys.* **45**, N 3, pp. 902–921.

[147] Khrennikov, A. (2004). Växjö interpretation-2003: Realism of contexts, in *Proc. Int. Conf. Quantum Theory: Reconsideration of Foundations, Ser. Math. Modelling* **10** (Växjö Univ. Press), pp. 323–338.

[148] Khrennikov, A. (2004). *Information Dynamics in Cognitive, Psychological, Social, and Anomalous Phenomena*, Fundamental Theories of Physics (Kluwer, Dordrecht).

[149] Khrennikov, A. (2004). Representation of cognitive information with the aid of probability distributions on the space of neuronal trajectories, *Proc. Steklov Inst. Math.* **245**, pp. 117–134.

[150] Khrennikov, A. (2004). Probabilistic pathway representation of cognitive information, *J. Theor. Biology* **231**, pp. 597–613.

[151] Khrennikov, A. (2004). On quantum-like probabilistic structure of mental information, *Open Systems and Information Dynamics* **11** (3), pp. 267–275.

[152] Khrennikov, A. (2005). Reconstruction of quantum theory on the basis of the formula of total probability, in *Foundations of Probability and Physics— 3, AIP Conf. Proc.* **750** (Amer. Inst. Phys., Melville, NY), pp. 187–218.

[153] Khrennikov, A. (2005). The principle of supplementarity: A contextual probabilistic viewpoint to complementarity, the interference of probabilities, and the incompatibility of variables in quantum mechanics, *Found. Phys.* **35**, N 10, pp. 1655–1693.

[154] Khrennikov, A. (2006). Quantum-like brain: Interference of minds, *BioSystems* **84**, pp. 225–241.

[155] Khrennikov, A. (2008). Bell-Boole inequality: Nonlocality or probabilistic incompatibility of random variables?, *Entropy* **10**, pp. 19–32.

[156] Khrennikov, A. (2009). *Interpretations of Probability*, 2nd edn. (de Gruyter, Berlin).

[157] Khrennikov, A. (2009). Bell's inequality, Einstein, Podolsky, Rosen arguments and von Neumann's projection postulate, *Laser Phys.* **19**, N 2, pp. 346–356.

[158] Khrennikov, A. (2009). EPR "paradox", projection postulate, time synchronization "nonlocality", *Int. J. Quant. Inf.* **7**, N 1, pp. 71–81.

[159] Khrennikov, A. (2008). The role of von Neumann and Lüders postulates in the Einstein, Podolsky, and Rosen considerations: Comparing measurements with degenerate and nondegenerate spectra, *J. Math. Phys.* **49**, N 5, 052102.

[160] Khrennikov, A. (2008). Analysis of the role of von Neumann's projection postulate in the canonical scheme of quantum teleportation, *J. Russian Laser Research* **29**, N 3, pp. 296–301.

[161] Khrennikov, A. (2009). *Contextual Approach to Quantum Formalism*, (Springer, Berlin-Heidelberg-New York)

[162] Khrennikov, A. (2010). Modelling of psychological behavior on the basis of ultrametric mental space: Encoding of categories by balls, *P-Adic Numbers, Ultrametric Analysis, and Applications* **2**, N 1, pp. 1–20.

[163] Khrennikov, A. (2010). *Ubiquitous quantum structure: from psychology to finances* (Springer, Berlin-Heidelberg-New York).

[164] Khrennikov, A. (2011). Quantum-like model of processing of information in the brain based on classical electromagnetic field, *Biosystems* **105(3)**, pp. 250–262.

[165] Khrennikov, A. (2014). *Introduction to foundations of probability and randomness (for students in physics)*, Lectures given at the Institute of Quantum Optics and Quantum Information, Austrian Academy of Science, Lecture-1: Kolmogorov and von Mises, arXiv:1410.5773 [quant-ph].

[166] Khrennikov, A., Basieva, I., Dzhafarov E. N. and Busemeyer, J. R. (2014). Quantum Models for Psychological Measurements: An Unsolved Problem, *PLoS ONE* **9**, e110909.

[167] Khrennikov, A. (2014). *Beyond Quantum* (Pan Stanford Publishing, Singapore).

[168] Khrennikov, A. (2015). CHSH inequality: Quantum probabilities as classical conditional probabilities, *Found. Phys.* **45**, pp. 711–725.

[169] Khrennikov, A., Ramelow, S., Ursin, R., Wittmann, B., Kofler, J. and Basieva, I. (2014). On the equivalence of the Clauser-Horne and Eberhard inequality based tests, *Phys. Scr.* **T163**, 014019.

[170] Khrennikov, A. (2015). Two-slit experiment: Classical and quantum probabilities. *Phys. Scr.* **90**, pp. 074017-1–074017-4.

[171] Khrennikov, A. (2015). Unuploaded experiments have no result, arXiv:1505.04293 [quant-ph].

[172] Khrennikova, P. (2012). Evolution of quantum-like modeling in decision making processes, *AIP Conf. Proc.* **1508**, pp. 108–112.

[173] Khrennikova, P., Haven, E. and Khrennikov, A. (2014). An application of the theory of open quantum systems to model the dynamics of party governance in the US political system, *Int. J. Theor. Phys.* **53**, pp. 1346–1360.

[174] Kofler, J. and Zeilinger, A. (2010). Quantum information and randomness, *European Rev.* **18**, N 4, pp. 469–480.

[175] Kolmolgorov, A. N. (1930). Sur la loi forte des grands nombre, *C.R. Acad. Sci.* **191** (Paris), pp. 910–912.

[176] Kolmolgorov, A. N. (1936). *The Basic Notions of Probability Theory* (FIAZ, Moscow), in Russian.

[177] Kolmolgoroff, A. N. (1933). *Grundbegriffe der Wahrscheinlichkeitsrechnung* (Springer-Verlag, Berlin).

[178] Kolmolgorov, A. N. (1956). *Foundations of the Probability Theory* (Chelsea Publ. Comp., New York).

[179] Kolmolgorov, A. N. (1974). *The Basic Notions of Probability Theory* (Nauka, Moscow), in Russian.

[180] Kolmolgorov, A. N. (1963). On tables of random numbers, *Sankhya Ser. A* **25**, pp. 369–375.

[181] Kolmolgorov, A. N. (1998). On tables of random numbers, *Theor. Comp. Sc.* **207 (2)**, pp. 387–395.

[182] Kolmolgorov, A. N. (1965). Three approaches to the quantitative definition of information, *Problems Inform. Transmition* **1**, pp. 1–7.

[183] Kolmolgorov, A. N. (1968). Logical basis for information theory and probability theory, *IEEE Trans. IT* **14**, pp. 662–664.

[184] Kolmolgorov, A. N. and Fomin, S.V. (1975). *Introductory Real Analysis* (Dover Publ.).

[185] Kracklauer, A. F. (2007). What do correlations tell us about photons?, in Ch. Roychoudhuri, A. F. Kracklauer and K. Creath (eds.), *The Nature of Light: What Are Photons?*, *Proc. SPIE* **6664**, 66640H.

[186] Kracklauer, A. F., Rangacharyulu, Ch., Roychoudhuri, Ch., Brooks, H. J., Carroll, J. and Khrennikov, A. (2009). Is indivisible single photon really essential for quantum communications, computing and encryption? in Ch. Roychoudhuri, A. F. Kracklauer and A. Yu. Khrennikov (eds.), *The Nature of Light: What are Photons? III, Proc. SPIE* **7421**, 74210Y.

[187] Kujala, J. V. and Dzhafarov, E. N. (2016). Probabilistic contextuality in EPR/Bohm-type systems with signaling allowed, in E. N. Dzhafarov et al. (eds.), *Contextuality from Quantum Physics to Psychology* (New Jersey: World Scientific), pp. 287–308.

[188] Kupczynski, M. (1987). Pitowsky model and complementarity, *Phys. Lett. A* **121**, pp. 51–53.

[189] Kupczynski, M. (1987). Bertrand's paradox and Bell's inequalities, *Phys. Lett. A* **121**, pp. 205–207.

[190] Kupczynski, M. (2007). EPR paradox, locality and completeness of quantum theory, in G. Adenier, A. Khrennikov, P. Lahti, V. Man'ko and T. Nieuwenhuizen (eds.), *Quantum Theory Reconsideration of Foundations-4, AIP Conference Proceedings* **962** (American Institute of Physics, Melville, NY), pp. 274–285.

[191] Kupczynski, M. (2009). Is quantum theory predictably complete?, *Phys. Scr.* **T135**, 014005.

[192] Kupczynski, M. (2011). Time Series, Stochastic Processes and Completeness of Quantum Theory Advances Quantum Theory, in G. Jaeger, A. Khrennikov, M. Schlosshauer, G. Weihs (eds.), *AIP Conference Proceedings* **1327** (American Institute of Physics, Melville, NY), pp. 394–400.

[193] Lande, A. (1965). *New Foundations of Quantum Mechanics* (Cambridge Univ. Press, Cambridge).

[194] Lüders, G. (1951). Uber die Zustandsanderung durch den Messprozess, *Ann. Phys.* **8** (Leipzig), pp. 322–328.

[195] Mackey, G. W. (1963). *Mathematical Foundations of Quantum Mechanics* (W. A. Benjamin Inc., New York).

[196] Margenau, H. (1958). Philosophical problems concerning the meaning of measurement in physics. *Phil. Sc.* **25**, pp. 23–33.

[197] Margenau, H. (1949). Reality in quantum mechanics, *Phil. Sc.* **16**, pp. 287–302.

[198] Martin-Löf, P. (1966). On the concept of random sequence, *Theory of Probability Appl.* **11**, pp. 177–179.

[199] Martin-Löf, P. (1966). The definition of random sequence, *Inform. Contr.* **9**, pp. 602–619.

[200] Martin-Löf, P. (1971). Complexity oscillations in infinite binary sequences, *Z. Wahrscheinlichkeitstheorie Verw.* **19**, pp. 225–230.

[201] Mermin, N. D. (2002). Whose knowledge?, in A. Yu. Khrennikov (ed.), *Quantum Theory: Reconsideration of Foundations, Ser. Math. Model* **2** (Växjö University Press, Växjö), pp. 261–270.

[202] Mermin, N. D. (2014). Why QBism is not the Copenhagen interpretation and what John Bell might have thought of it, Preprint: arXiv:1409.2454v1 [quant-ph].

[203] Michielsen, K., De Raedt, H. and Hess, K. (2011). Boole and Bell inequality, in G. Jaeger, A. Khrennikov, M. Schlosshauer, G. Weihs (eds.), *Advances in Quantum Theory, AIP Conference Proceedings*, Vol. 1327 (Melville–New York), pp. 429–433.

[204] Mizuguchi, T. and Suwashita, M. (2006). Dynamics of coin tossing, *Prog. Theor. Phys. Suppl.* **161**, p. 274.

[205] Mückenheim, W. (1986). A review on extended probabilities, *Phys. Rep.* **133**, pp. 338-401.

[206] Nieuwenhuizen, Th. M. (2009). Where Bell went wrong?, in L. Accardi, G. Adenier, C. A. Fuchs, G. Jaeger, A. Yu. Khrennikov, J.-A. Larsson and S. Stenholm (eds.), *AIP Conf. Proc., Foundations of Probability and Physics - 5* **1101** (Am. Inst. Phys., Melville, NY), pp. 127–133.

[207] Nyman, P. and Basieva, I. (2011). Representation of probabilistic data by complex probability amplitudes; the case of triplevalued observables, in *Advances in Quantum Theory, AIP Conf. Proc.* **1327**, pp. 439–449.

[208] Nyman, P.and Basieva, I. (2011). Quantum-like representation algorithm for trichotomous observables, *Int. J. Theor. Phys.* **50**, pp. 3864–3881.

[209] Oas, G., de Barros, J. Acacio and Carvalhaes, C. (2014). Exploring non-

signalling polytopes with negative probability, *Phys. Scr.* **T163**, 014034.

[210] Ohya, M. and Volovich, I. (2011). *Mathematical Foundations of Quantum Information and Computation and its Applications to Nano- and Biosystems* (Springer, Heidelberg-Berlin-New York).

[211] Ozawa, M. (1984). Quantum measuring processes of continuous observables, *J. Math. Phys.* **25**, p. 79.

[212] Ozawa, M. (1997). An operational approach to quantum state reduction, *Ann. Phys. (N.Y.)* **259**, pp. 121–137.

[213] Ozawa, M. (2003). Universally valid reformulation of the Heisenberg uncertainty principle on noise and disturbance in measurement, *Phys. Rev. A* **67**, 042105-(1–6).

[214] Pais, A. (1983). *Subtle Is the Lord: The Science and the Life of Albert Einstein* (Oxford Univ. Press, Oxford).

[215] Pauli, W. (1994). *Writings on Physics and Philosophy* (Springer, Berlin).

[216] Penrose, R. (1989). *The Emperor's New Mind* (Oxford Univ. Press, New York).

[217] Pironio, S., Acin, A., Massar, S., Boyer de la Giroday, A., Matsukevich, D. N., Maunz, P., Olmschenk, S., Hayes, D., Luo, L., Manning, T. A. and Monroe, C. (2010). Random Numbers Certified by Bell's Theorem, *Nature* **464**, p. 1021.

[218] Pitowsky, I. (1982). Resolution of the Einstein-Podolsky-Rosen and Bell paradoxes, *Phys. Rev. Lett.* **48**, 1299; *Erratum Phys. Rev. Lett.* **48**, 1768.

[219] Plotnitsky, A. (2001). Reading Bohr: Complementarity, epistemology, entanglement, and decoherence, in A. Gonis and P. Turchi (eds.), *Proc. NATO Workshop "Decoherence and its Implications for Quantum Computations"* (IOS Press, Amsterdam), pp. 3–37.

[220] Plotnitsky, A. (2006). *Reading Bohr: Physics and Philosophy* (Springer, Heidelberg-Berlin-New York).

[221] Plotnitsky, A. (2009). *Epistemology and Probability: Bohr, Heisenberg, Schrdinger and the Nature of Quantum-Theoretical Thinking* (Springer, Berlin and New York).

[222] Plotnitsky, A. (2012). *Niels Bohr and Complementarity: An Introduction* (Springer, Berlin and New York).

[223] Plotnitsky, A.and Khrennikov, A. (2015). Reality without realism: On the ontological and epistemological architecture of quantum mechanics, *Found. Phys.* **45**, N 10, pp. 269–1300.

[224] Primas, H. (1990). Mathematical and philosophical questions in the theory of open and macroscopic quantum systems, in A. I. Miller (ed.) *In Sixty-Two Years of Uncertainty* (Plenum, New York), pp. 233–257.

[225] Ramsey, F. P. (1931). "Truth and Probability," in R. B. Braithwaite (ed.), *The Foundations of Mathematics and other Logical Essays* (Harcourt, Brace and Company, New York), p. 156.

[226] Savage, L. J. (1954). *The Foundations of Statistics* (John Wiley & Sons, New York).

[227] Schnorr, C. P. (1971). Zufalligkeit und Wahrscheinlichkeit, *Lect. Notes in Math.* **218** (Springer-Verlag, Berlin).

[228] Schrödinger, E. (1926). Quantisierung als Eigenwertproblem, I, *Ann. Physik* **79**, pp. 361–376.

[229] Schrödinger, E. (1926). Quantisierung als Eigenwertproblem, II, *Ann. Physik* **79**, pp. 489–527.

[230] Schrödinger, E. (1926). An undulatory theory of the mechanics of atoms and molecules, *Phys. Rev.* **28**, pp. 1049–1070.

[231] Schrödinger, E. (1935). Die gegenwrtige Situation in der Quantenmechanik, *Naturwissenschaften* **23**, pp. 807–812; 823–828; 844–849.

[232] Shafer, G. (1985). Moral certainty, in S. Kotz and N. L. Johnson (eds.), *Encyclopedia of Statistical Sciences* **5** (Wiley), pp. 623–624.

[233] Shiryaev, A. N. (1984). *Probability* (Springer, New York-Berlin-Heidelberg).

[234] Siegmund-Schultze, R. (2010). Sets versus trial sequences, Hausdorff versus von Mises: "Pure" mathematics prevails in the foundations of probability around 1920, *Historia Mathematica* **37**, pp. 204–241.

[235] Singpurwalla, N. and Smith, R. (2006). A conversation with B. V. Gnedenko, *Reliability: Theory & Applications*, No 1, pp. 82–88.

[236] Smolaynov, O. G. and Khrennikov, A. Yu. (1985). Central limit theorem for generalized measures on infinite-dimensional space, *Dokl. Akad. Nauk USSR, ser. Mat.* **281**, pp. 279–283.

[237] Solomonoff, R. J. (1960). A preliminary report on a general theory of inductive inference, *Report V-131* (Cambridge, Ma.: Zator Co.).

[238] Solomonoff, R. J. (1964). A formal theory of inductive inference, *Information and Control* **7**, pp. 1–22.

[239] Strzalko, J., Grabski, J., Stefanski, A., Perlikowski, P. and Kapitaniak, T. (2010). Understanding coin-tossing, *Math. Intelligence* **32**, N 4, pp. 54–58.

[240] Sulyok, G., Sponar, S., Erhart, J., Badurek, G., Ozawa, M. and Hasegawa, Y. (2013). Violation of Heisenberg's error-disturbance uncertainty relation in neutron-spin measurements, *Phys. Rev. A* **88**, 022110.

[241] Tegmark, M. (2000). Importance of quantum decoherence in brain processes, *Phys. Rev. E* **61** (4), pp. 4194-4206.

[242] 't Hooft, G. (2011). The free-will postulate in quantum mechanics, *Herald of Russian Academy of Science* **81**, pp. 907–911, Preprint (2007). ArXiv: quant-ph/0701097v1.

[243] 't Hooft, G. (1999). *Quantum gravity as a dissipative deterministic system*, preprint ArXiv: gr-qc/9903084.

[244] 't Hooft, G. (2006). *The mathematical basis for deterministic quantum mechanics*, preprint ArXiv: quant-ph/0604008.

[245] Ville, J. (1939). *Etude critique de la notion de collective* (Gauthier-Villars, Paris).

[246] Vladimirov, V. S., Volovich, I. V. and Zelenov, E. I. (1994). *p-adic Analysis and Mathematical Physics* (World Scientific, Singapore).

[247] von Mises, R. (1919). Grundlagen der Wahrscheinlichkeitsrechnung, *Math. Z.* **5**, pp. 52–99.

[248] von Mises, R. (1957). *Probability, Statistics and Truth* (Macmillan, London).

[249] von Mises, R. (1964). *The mathematical theory of probability and statistics*

(Academic, London).

[250] von Neumann, J. (1955). *Mathematical Foundations of Quantum Mechanics* (Princeton Univ. Press, Princenton).

[251] Vorob'ev, N. N. (1962). Consistent families of measures and their extensions, *Theory of Probability and its Applications* **7**, pp. 147–162.

[252] Vulovic, V. Z. and Prange, R. E. (1986). Randomness of true coin toss, *Phys. Rev. A* **33**, p. 576.

[253] Wald, A. (1938). Die Widerspruchsfreiheit des Kollektivbegriffs in der Wahrscheinlichkeitsrechnung, *Ergebnisse eines Math. Kolloquiums* **8**, pp. 38–72.

[254] Weihs, G. (1999). *Ein Experiment zum Test der Bellschen Ungleichung unter Einsteinscher Lokalitat* (University of Vienna Press, Vienna).

[255] Weihs, G., Jennewein, T., Simon, C., Weinfurter, R. and Zeilinger, A. (1998). Violation of Bell's inequality under strict Einstein locality conditions, *Phys. Rev. Lett.* **81**, pp. 5039–5043.

[256] Weihs, G. (2007). A test of Bell's inequality with spacelike separation, in: G. Adenier, C. Fuchs and A. Yu. Khrennikov (eds.), *Foundations of Probability and Physics-4, Ser. Conference Proceedings* **889** (American Institute of Physics, Melville, NY), pp. 250–262.

[257] Wheeler, John A. (1990). "Information, physics, quantum: The search for links", in Zurek, Wojciech Hubert *Complexity, Entropy, and the Physics of Information* (Redwood City, California: Addison-Wesley).

[258] Zeilinger, A. (1999). A foundational principle for quantum mechanics, *Foundations of Physics* **29**, pp. 631–641.

[259] Zeilinger, A. (2010). *Dance of the Photons: From Einstein to Quantum Teleportation* (Farrar, Straus and Giroux, New-York).

[260] Zhang, Ya., Glancy, S. and Knill, E. (2011). Asymptotically optimal data analysis for rejecting local realism, *Phys. Rev. A* **84**, 062118.

Index

Printed in the United States
By Bookmasters